APPLIED ECOLOGY

Other Titles in the Project

Biology   Martin Rowland
Applied Genetics   Geoff Hayward
Micro-organisms and Biotechnology   Jane Taylor
Biochemistry and Molecular Biology   Moira Sheehan

Physics   Robert Hutchings
Telecommunications   John Allen
Medical Physics   Martin Hollins
Nuclear Physics   David Sang
Energy   David Sang and Robert Hutchings

Chemistry   Ken Gadd and Steve Gurr

UNIVERSITY OF BATH • SCIENCE 16-19

Project Director: J. J. Thompson, CBE

APPLIED ECOLOGY

GEOFF HAYWARD

Nelson

Thomas Nelson and Sons Ltd
Nelson House, Mayfield Road
Walton-on-Thames, Surrey
KT12 5PL UK

51 York Place
Edinburgh
EH1 3JD UK

Thomas Nelson (Hong Kong) Ltd
Toppan Building 10/F
22A Westlands Road
Quarry Bay, Hong Kong

Thomas Nelson Australia
102 Dodds Street
South Melbourne
Victoria 3205 Australia

Nelson Canada
1120 Birchmount Road
Scarborough, Ontario
M1K 5G4 Canada

© Geoff Hayward 1992,

First published by Thomas Nelson and Sons Ltd 1992

ISBN 0-17-448187-X
NPN 9 8 7 6 5 4 3 2 1

Printed in Great Britain by
Ebenezer Baylis & Son Ltd,
The Trinity Press, Worcester, and London

Contents

The Project: an introduction   vii

How to use this book   viii

Theme 1:   The ecology of marine fisheries

Chapter 1	**The natural history of the sea**	2
1.1	Marine habitats	2
1.2	Who's who in the sea	4
1.3	Food webs and food chains	7
1.4	The movement of water in the sea	10
1.5	Thermoclines	14
Chapter 2	**Marine productivity**	17
2.1	Primary production in the sea	17
2.2	Factors limiting marine primary productivity – light	20
2.3	Factors limiting marine primary productivity – nutrients	22
2.4	Factors limiting marine primary productivity – turbulence	24
2.5	Seasonal variation in marine primary productivity	27
2.6	Ecological efficiency of marine food chains	29
Chapter 3	**Fish and fisheries**	32
3.1	How fish live	32
3.2	Life history	33
3.3	Fish population dynamics	37
3.4	The fishing industry	41
Chapter 4	**Harvesting the sea 1: Fishery models**	47
4.1	Fish population dynamics	48
4.2	Harvesting the stock	50
4.3	Gathering fisheries data	51
4.4	Managing fisheries	53
4.5	Fixed quota and fixed effort harvesting	55
Chapter 5	**Harvesting the sea 2: The overfishing problem**	60
5.1	Recognising and preventing overfishing	61
5.2	Case study: the Peruvian anchoveta industry	62
5.3	Case study: the North Sea fisheries	65
5.4	Fish farming	67

Theme 2:   Agricultural ecology

Chapter 6	**Plant growth analysis**	74
6.1	The efficiency of photosynthesis	74
6.2	Measuring plant performance	76
6.3	Growth analysis of individual plants	79
6.4	Growth analysis of populations and communities of plants	81
6.5	Monitoring crop quality	86
Chapter 7	**Climatic factors affecting productivity**	89
7.1	The principle of limiting factors	90
7.2	Agriculture and climate – an introduction	91
7.3	Temperature and plant growth	96
7.4	Light	99
7.5	Wind	103

Chapter 8	**Soil and edaphic factors**	106
8.1	The nature of edaphic factors	106
8.2	The physical structure of soils	110
8.3	Soil porosity	113
8.4	Major soil types	116
Chapter 9	**Soil management**	119
9.1	Tilth and tillage	120
9.2	Liming	122
9.3	Fertilisers	125
9.4	Manuring	129
9.5	Soil water management	132
Chapter 10	**Pests – biotic factors affecting crop productivity**	136
10.1	What is a pest?	136
10.2	Weeds	139
10.3	Animal pests	143
10.4	Plant diseases	144
Chapter 11	**Pest control**	148
11.1	Protecting crops – cultural methods	148
11.2	Chemical control – herbicides and fungicides	151
11.3	Insecticides	154
11.4	Biological control	159
11.5	Integrated pest management	161

Theme 3: Freshwater pollution

Chapter 12	**Dirty water**	166
12.1	What is pollution?	166
12.2	Types, sources and effects of freshwater pollutants	168
12.3	The fate of freshwater pollutants	173
12.4	Case study: sewage	177
Chapter 13	**Organic pollution**	179
13.1	Sources of biodegradable organic pollutants	179
13.2	Using up oxygen	180
13.3	The effect of BOPs on plants and animals	183
Chapter 14	**Acid rain, eutrophication and toxic pollution**	190
14.1	Acid rain	190
14.2	Eutrophication	195
14.3	Toxic pollution	199
Chapter 15	**Monitoring water quality**	204
15.1	Routine testing of water	204
15.2	Measuring oxygen demand	206
15.3	Biological monitoring	210
Appendix A	**Answers to analysis exercises**	217
Appendix B	**Answers to questions**	221
	Index	229

The Project: an introduction

The University of Bath Science 16–19 Science Project grew out of reappraisal of how far sixth form science had travelled during a period of unprecedented curriculum reform and an attempt to evaluate future development. Changes were occurring both within the constitution of 16–19 syllabuses themselves and as a result of external pressures from 16+ and below; syllabus redefinition (starting with the common cores), the introduction of AS-level and its academic recognition, the originally optimistic outcome to the Higginson enquiry; new emphasis on skills and processes, and the balance of continuous and final assessment at GCSE level.

This activity offered fertile ground for the School of Education at the University of Bath and a major publisher to join forces with a team of science teachers, drawn from a wide spectrum of educational experience, to create a flexible curriculum model and then develop resources to fit it. The group addressed the task of satisfying these requirements:

- the new syllabus and examination demands of A and AS-level courses;
- the provision of materials suitable for both the core and options parts of syllabuses;
- the striking of an appropriate balance of opportunities for students to acquire knowledge and understanding, develop skills and concepts, and to appreciate the applications and implications of science;
- the encouragement of a degree of independent learning through highly interactive texts;
- the satisfaction of the needs of a wide range of students at this level.

Some of these objectives were easier to achieve than others. Relationships to still evolving syllabuses demand the most rigorous analysis and a sense of vision – and optimism – regarding their eventual destination. Original assumptions about AS-level, for example, as a distinct though complementary sibling to A-level, needed to be revised.

The Project, though, always regarded itself as more than a provider of materials, important as this is, and concerned itself equally with the process of provision – how material can best be written and shaped to meet the requirements of the educational market-place. This aim found expression in two principal forms: the idea of secondment at the University and the extensive trialling of early material in schools and colleges.

Most authors enjoyed a period of secondment from teaching, which allowed them to reflect and write more strategically (and, particularly so in a supportive academic environment) but, equally, to engage with each other in wrestling with the issues in question.

The Project saw in the trialling a crucial test for the acceptance of its ideas and their execution. Over one hundred institutions and one thousand students participated, and responses were invited from teachers and pupils alike. The reactions generally confirmed the soundness of the model and allowed for more scrupulous textual housekeeping, as details of confusion, ambiguity or plain misunderstanding were revised and reordered.

The test of all teaching must be in the quality of the learning, and the proof of these resources will be in the understanding and ease of accessibility which they generate. The Project, ultimately, is both a collection of materials and a message of faith in the science curriculum of the future.

J.J. Thompson
January 1992

How to use this book

Ecology is an area of intense scientific research and one where the fundamental principles uncovered can quickly be applied to managing our environment. This book aims to help you learn about these fundamental ecological principles and how they are applied in three areas – marine biology and fisheries management, agricultural ecology and the ecology of freshwater pollution. In addition the book also looks at the social and economic problems associated with managing our environment.

Applied Ecology is written for A- or AS-level courses in biology and in particular for options in ecology. Ecology is a subject students often find difficult and a primary objective has been to present information in an accessible and interesting way. The book assumes and understanding of biology up to GCSE level and some topics also require some understanding of topics which are covered in A-level biology courses.

The book is divided into three themes and can be used in a number of different ways. For example, you could use it as a course book working your way through all the chapters in the order in which they are given. Alternatively, you might only need to study some or the chapters or you may wish to use the questions and summary assignments to help in revision.

The best way to study any subject is to take some responsibility for your own learning. This means, for example, ensuring that you make a good set of notes and answer the questions as you come to them so that you can assess your progress. The book is designed to help you organise your studies and on the page opposite you will find information about the book which will help you get the best out of it.

Reading is, on its own, usually too passive to promote effective thinking and learning and the questions within the chapters are, therefore, an important feature of this book. They are intended to help you understand what you have just read. You should write down the answers to the questions as you come to them and then check the answers with those at the back of the book or with your teacher. If you do not understand a question or an answer, make a note of it and discuss it with your teacher at the earliest opportunity.

Throughout the book you are encouraged to use and develop the skills essential for success at A-level. In particular the Analysis boxes provide longer exercises involving data handling, graphical, observational and problem-solving skills. In addition, there are some more extended project suggestions where you can work together with other members of your class.

Ecology is a very active and exciting area of biological research which offers a multitude of careers. I hope this book will not only enable you to learn the material you need to pass an exam, but also encourage you to investigate whether you would like to work in this area.

Learning objectives

These are given at the beginning of each chapter and they outline what you should gain from the chapter. They are statements of attainment and often link closely to statements in a course syllabus. Learning objectives can help you make notes for revision, especially if used in conjunction with the summary assignments at the end of the chapter, as well as for checking progress.

Questions

In-text questions occur at points when you should consolidate what you have just learned, or prepare for what is to follow, by thinking along the lines required by the question. Some questions can, therefore, be answered from the material covered in the previous section, others may require additional thought or information. Answers to questions are at the end of the book.

Analysis boxes

These provide the opportunity to develop essential skills in problem solving, graphicacy and data analysis, all essential for success at A-level. They will also help you to check your own learning and understanding. Answers to most analyses are given at the end of the book.

Summary assignments

These require you to think about the key points in the chapter and so can be used to build up your understanding of the topic. They can, therefore, be used to compile your own revision notes, or at the end of the course to test your understanding of a chapter.

Acknowledgements

The author and publishers wish to thank the following who have kindly given permission for the use of copyright material:

Joint Matriculation Board for past examination questions;
Ministry of Agriculture, Fisheries and Food for Fig 3.11 from part of Fig 9.34 in *Fishery Investigations Series 2, Vol XIII, No 4, 1934*;
The Open University for Table 3.1 from Table 6, p. 43, S334, Unit 9;

The authors and publishers wish to acknowledge, with thanks, the following photographic sources:

AEA Harwell Technology *p 205 right*; Barnaby's Picture Library *pp 65, 68 right, 93, 117, 120 top, 194*; Biofotos *pp 33, 69, 84 bottom, 165 left*; British Coal *p 209*; J Allan Cash Photo Library *pp 68 left, 84 top, 165 right*; CTC Publicity *pp 158, 182*; Henry Doubleday Research Association *p 150*; Environmental Picture Library *pp 167 left, 196*; Farmers Weekly *pp 74, 121, 125 bottom, 143 bottom, 146, 149*; Fisons *p 209 left*; Geoscience Features *pp 107 top, 120 bottom*; Glasshouse Crops Research Institute *p 159*; Greenpeace *pp 167 centre, 202, 204*; The Guardian/Dennis Thorpe *p 47*; Holt Studios *pp 73 101, 107 bottom, 124 left & right, 125 top, 130, 136, 138, left, 143 top, 152, 156, 179*; MAFF, Directorate of Fisheries Research *pp 36, 38, 52, 66*; National Rivers Authority *p 211*; New Scientist *p 70*; N T Nicoll *pp 5, 7*; Oxford Scientific Films *p 138 right*; Popperfoto *p 103*; USDA Soil Conservation Service *p 198*; Science Photo Library *p 92* South American Pictures *p 1*; Survival Anglia *p 57*; Topham *p 62*; Water Research Council *p 186*; Wildlife Matters *pp 167 right, 170*.

Theme 1

THE ECOLOGY OF MARINE FISHERIES

Fish and shellfish supply about 6% of all protein consumed by human beings and 24% of protein included in animal feeds – considerably more than beef, twice as much as eggs and three times as much as poultry. Fish and shellfish are the major source of animal protein, iron and iodine for more than half the world's people, especially in Asia and Africa. As the population of the world increases there will be an increasing demand for more and more fish to be taken from the sea. However, such an increase in exploitation needs to be managed carefully so that stocks are not overfished. The production of such management policies and their application are discussed in this theme.

To enable you to understand the ecology of marine fisheries you need some background information about marine biology. In particular you need to know:

* the types of marine habitats and the sorts of organisms that live in those habitats;

* how marine organisms are linked by food chains and organised into food webs;

* how water movements affect marine organisms;

* the factors which control the productivity of marine organisms.

The first three of these points are dealt with in Chapter 1 while the fourth is the subject of Chapter 2. As these two chapters are closely linked, and are necessary to understand and complete Chapters 3 to 5 successfully, you must ensure that you work through these chapters thoroughly and in order.

Fifty tonnes of anchovies were obtained from this single haul off the coast of Peru in the mid-1960s using this purse seine net. Note the use of a vacuum pipe to suck up these small fish.

Chapter **1**

THE NATURAL HISTORY OF THE SEA

LEARNING OBJECTIVES

After completing the work in this chapter you will be able to:

1. divide the marine ecosystem into four basic habitat types: benthic, pelagic, neritic and oceanic;

2. give an ecological account of the nature of planktonic organisms;

3. compare and contrast pelagic and benthic food webs;

4. discuss the importance of microorganisms in marine food webs and nutrient cycling;

5. compare the effects of water movements and thermal stratification (**thermoclines**) on the structure of the water column.

1.1 MARINE HABITATS

Essentially, there are two types of marine organism: those that live on or in the sea bed, **benthic organisms**, and those that live surrounded by water, **pelagic organisms**. The **benthic division** of marine habitats is shown in red in Fig 1.1 while the **pelagic division** is shown in grey. Note that the pelagic division is further subdivided into the **neritic** and **oceanic provinces**. Neritic waters are those that overlay the continental shelf and are relatively shallow (100–200 m) compared with the deeper oceanic waters. As we will see these two types of water can have very different properties which, in turn, lead to differences in their productivity and hence their importance as fishing areas.

The benthic division

Imagine you are standing at point A in Fig 1.1. You start to walk out into the sea along the sea bottom. You pass through the **littoral zone**, the narrow strip of land which is covered by the sea at high tide but uncovered at low tide, and into the **sublittoral zone**. At this stage of your journey you would be walking along the gently sloping **continental shelf**. The extent of the continental shelf varies but if you keep walking you will eventually reach the **continental edge**. At this point the slope of the sea bottom changes dramatically as you begin to descend the continental slope and enter the **bathybenthic zone**. Eventually, you will reach the bottom of the continental slope and encounter the more gently sloping **continental rise**. By now, at a depth of over 4000 m, you will be in the **abyssobenthic zone**. The abyssal plain is essentially flat, of an enormous extent and is covered with a thick layer of sediment.

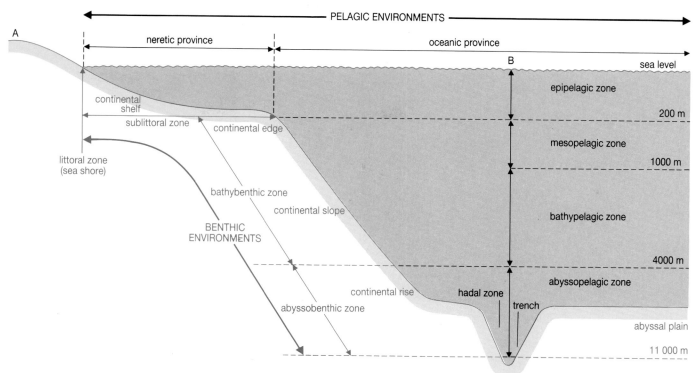

Fig 1.1 The structure of the marine environment. Note the two major environments and the divisions within those environments.

The pelagic division

The pelagic division is divided into four layers or zones.

1. The **euphotic** or **epipelagic zone** is that part of the pelagic division which is well illuminated. Its lower boundary varies in depth according to the clarity of the water but is usually around 100–200 m in oceanic waters, although it may be much less in neritic waters close to the coast where the water clarity is reduced by the growth of plankton and suspended material washed into the sea by rivers and brought up from the bottom of the sea by wave action. In addition to light there can also be a sharp change in temperature in this zone from warm water near the surface to very cold water lower down. There may also be large seasonal variations in the temperature of surface water in temperate regions.

2. The **mesopelagic zone**, lying between 200 and 1000 m, is virtually dark with only a gradual change in temperature with depth and little seasonal variation. This zone often contains the oxygen minimum layer and the maximum concentrations of nutrients like nitrate and phosphate (Fig 1.2).

3. Below 1000 m, in the **bathypelagic zone**, there is virtually no light with the exception of that produced by **bioluminescence**, the temperature is low and essentially constant (0.5–2.0°C) and the pressure is very great.

4. Finally, there are the deepest parts of the pelagic division, the **abyssopelagic zone** and the **hadal zone** of the great oceanic trenches, like the Marianas trench, the bottom of which, in the Challenges Deep, is 10 863 m below sea level.

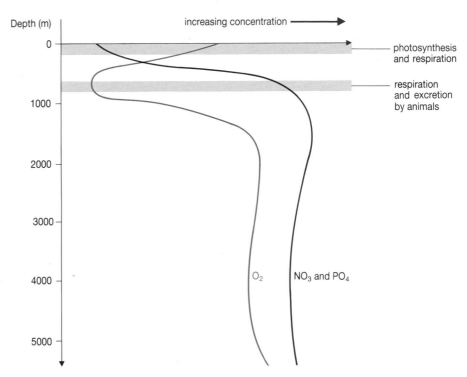

Fig 1.2 Oxygen, nitrate and phosphate concentration changes with depth. These changes are related to the rates of photosynthesis, respiration and excretion at different depths.

QUESTIONS

1.1 The air pressure at the sea surface is about 100 kPa. Each 10 m of sea water increases this pressure by a further 100 kPa. With reference to Fig 1.1 calculate the pressure at
   (a) the bottom of the epipelagic zone
   (b) the top of the abyssopelagic zone
   (c) on the abyssal plain which has an average depth of 4000 m
   (d) at the bottom of the Marianas trench.

1.2 Discuss with your friends and tutor the sort of adaptations you might expect to be shown by animals and plants which inhabit
   (a) the epipelagic zone
   (b) the abyssopelagic zone.

1.2 WHO'S WHO IN THE SEA

Marine organisms can be split into different categories depending on where they live and on their powers of movement. Some of these groups are defined in Table 1.1. In this and the next chapter we will be mainly concerned with the plankton so this group is considered in more detail below.

The plankton

Plankton are usually collected by means of nets or filters. Since these have a standard mesh or pore size, plankton are often classified on the basis of size. The main categories are shown in Table 1.2, which indicates that the size-based classification also corresponds to a systematic one.

For our purposes it is better to adopt an ecological classification of the plankton rather than a taxonomic one. This allows us to group together organisms which do the same job in marine ecosystems even though they come from different taxonomic groups. For example, some of the eukaryotic single-celled algae, which form the **phytoplankton**, can be

Table 1.1 A glossary of terms used in describing marine organisms.

Term	Description
plankton	free-floating organisms with limited power of locomotion
phytoplankton	the plant, primarily algal, component of plankton
zooplankton	animal component of plankton
bacterioplankton	planktonic bacteria
holoplankton	permanent members of the plankton
meroplankton	temporary members of the plankton, e.g. many larval stages of benthic organisms
neuston	small planktonic organisms inhabiting the ultra-thin sea surface layer
nekton	organisms capable of swimming against water currents, e.g. fish, squid, whales

Table 1.2 Classification of plankton by size.

Type	Size range	Examples
femtoplankton	0.002–0.2 µm	viruses
picoplankton	0.2–2.0 µm	bacteria
nanoplankton	2.0–20 µm	small autotrophic flagellates
microplankton	20–200 µm	protozoans, diatoms, dinoflagellates
mesoplankton	0.2–20 mm	copepods
macroplankton	2.0–20 cm	krill, arrow worms
megaplankton	0.2–2.0 m	large jellyfish

heterotrophic (those that obtain their energy and carbon from consuming other organisms or their products) rather than **autotrophic** (those that can convert, or fix, carbon dioxide into carbohydrates). We can avoid this taxonomic problem by adopting a functional classification of the plankton into autotrophic and heterotrophic species.

The autotrophic plankton

These are primarily single-celled organisms although they can occur as chains or small colonies of cells (Fig 1.3). The **photoautotrophs** (those that use sunlight as an energy source) are the principal primary producers in the sea and they can only grow when they have sufficient light. The **chemoautotrophs** can fix carbon using energy obtained from a variety of simple compounds and elements, including ammonia, methane, iron and sulphur. This process, **chemosynthesis**, can occur in the absence of light. Both types of organism absorb carbon dioxide, water and essential nutrients, like nitrate and phosphate, from the water surrounding them. In some environments, for example the depths of the Black Sea, the amount of carbon fixed by chemosynthesis can rival that fixed by photosynthesis, but as photosynthesis usually accounts for about 95% of the carbon fixed in the sea we will only consider photoautotrophs in detail here.

Traditionally, it has been the largest of the autotrophic plankton, the **microplankton**, which have been the most intensely investigated. The two most dominant members of this group are the diatoms and dinoflagellates.

Diatoms are characterised by the possession of a rigid cell wall impregnated with silica. The cell wall is divided into two overlapping halves called valves (Fig 1.4) which can take a variety of shapes from simple to complex which are characteristic of an individual species. Diatoms are usually yellowish brown in colour and may occur individually, e.g. *Coscinodiscus*, or as chains, e.g. *Chaetoceros*. In coastal waters and in

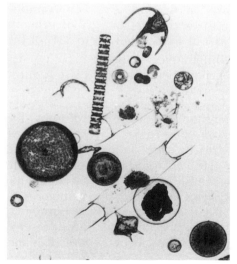

Fig 1.3 A collection of autotrophic plankton. Note the anchor shaped dinoflagellates and large round diatoms.

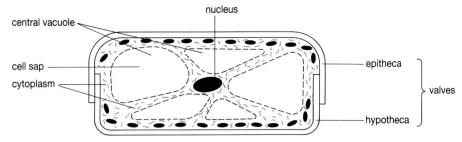

Fig 1.4 Diagrammatic section through a diatom. The two valves are made of silica.

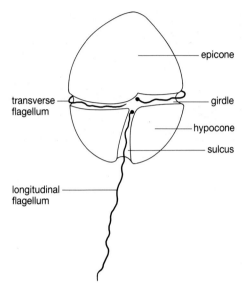

Fig 1.5 The position of the two flagella are easily seen in this dinoflagellate.

areas of upwelling (see Section 1.4), diatoms may contribute half the total biomass of the autotrophic plankton present in the sea.

Dinoflagellates (Fig 1.5) generally have the cell enclosed within cellulose plates of a precise pattern so, like the diatoms, they also have definite, often complicated shapes. They have a pair of flagella, one in a groove circling the body and the other projecting at right angles. They usually contain several chloroplasts and are usually yellow-green or yellow-brown although there are some colourless heterotrophic species. Some types, such as *Gonylaux*, produce toxins which, during **bloom** conditions, may cause 'red tides', which occasionally result in mass mortalities of fish and invertebrates. The consumption of shellfish, like mussels, which have been contaminated with these toxins may cause severe health problems, even death, in humans since the poisons are powerful neurotoxins.

Cyanobacteria or 'blue-green algae' are prokaryotic organisms, some of which are able to convert nitrogen directly into nitrate and so grow in water which has a low nitrate concentration. The importance of cyanobacteria in terms of marine primary production has been thought to be small but this has recently been called into question. Many small cells often pass through the filters designed to catch them and it is now known that very small coccoid (round) cyanobacteria, many with cell diameters of a micrometre (10^{-6} m) or less, are common and widespread. The contribution that these cells make to primary production remains to be assessed.

In addition to the tiny coccoid autotrophic cyanobacteria the **nanoplankton** and **picoplankton** also contain a large number of eukaryotic cells, both coccoid and flagellated. It is becoming increasingly obvious that these largely unknown organisms may be responsible for a substantial part of marine primary production particularly in tropical seas. For example, estimates suggest that picoplankton are responsible for 22% of primary production in the Celtic Sea off the coast of Ireland but up to 80% of the primary production in the waters around Hawaii.

Heterotrophic plankton

We can divide this group into **zooplankton** and **microheterotrophs** like bacteria and protozoa. The zooplankton constitute a diverse group of animals containing representatives of virtually every animal phylum. Some zooplankters spend their entire life in the plankton (**holoplankton**) while only the eggs and larvae of otherwise benthic animals may be planktonic (**meroplankton**). Copepods, such as *Calanus*, are important herbivores in temperate seas while the euphasiids (krill) are particularly important in upwelling areas around Antarctica. These herbivorous species are food for a wide range of carnivores, including comb jellies, arrow worms, fish and the giant baleen whales (Fig 1.6).

With the exception of the largely autotrophic cyanobacteria most marine bacteria are heterotrophic. In addition, there are large marine populations

THE NATURAL HISTORY OF THE SEA

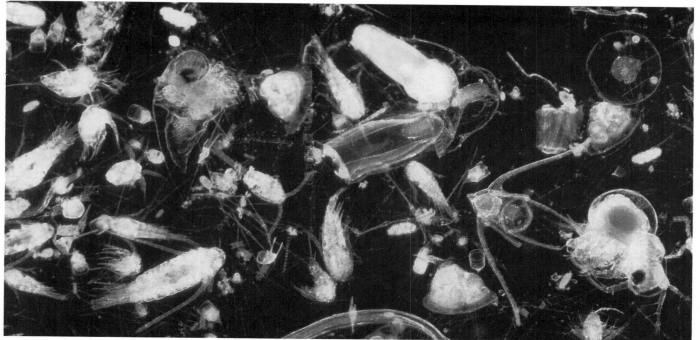

Fig 1.6 A collection of zooplankton. The majority of organisms in this photograph are crustaceans, primarily *Calanus*. Notice the round carnivorous jellyfish at the top right of the photograph and the echinoderm larva, the one with long "arms", below it.

of fungi and protozoa. The main processes in which these microorganisms are involved include:

- consuming and releasing much of the primary production of the autotrophic plankton as **dissolved organic matter (DOM)** or as dead **particulate organic matter (POM)**;
- colonisation and degradation of POM which releases nutrients;
- uptake of DOM with production of living and dead POM, thus forming food for other marine heterotrophs, especially protozoa and planktonic filter feeders, e.g. tunicates, and benthic detritivores.

1.3 FOOD WEBS AND FOOD CHAINS

The herring, a major pelagic food fish, has been the subject of much fisheries research, allowing the construction of the food web shown in Fig 1.7. At least three trophic levels can be distinguished.

1. Primary producers – organisms that use solar energy to convert inorganic compounds, e.g. carbon dioxide and water into organic compounds.

2. Herbivores or primary consumers – organisms that feed directly on the primary producers.

3. Carnivores, or secondary consumers – organisms that feed on the herbivores.

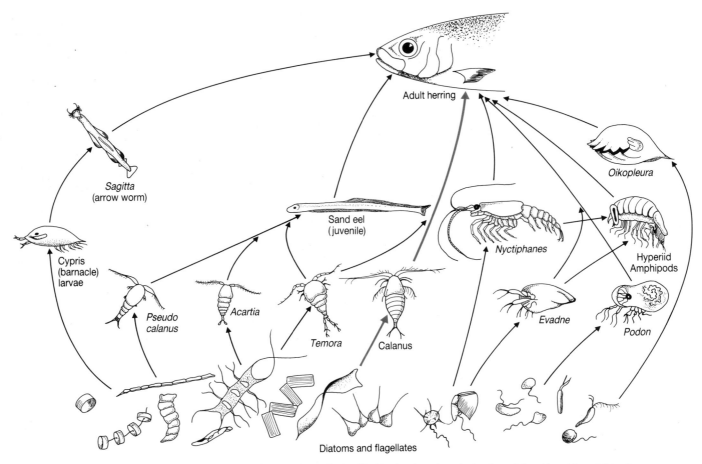

Fig 1.7 A greatly simplified food web to show the feeding relationships of the adult herring.

Pelagic food chains

Look again at Fig 1.7 but this time concentrate on the single food chain highlighted in red. Consider first the primary producers. To recap, these are small, mainly single-celled, free-floating organisms, including both algae and cyanobacteria, which are autotrophic. These cells synthesise organic substances, like carbohydrates, proteins and fats, which they then use to grow and divide. This growth, or **primary production**, is the first link in all marine food chains and the small autotrophic plankton are responsible for 99% of all marine primary production. The larger algae, the sea weeds, which are found attached to rocks in the littoral and sublittoral zones, account for only a tiny proportion of total marine primary production, even though they are much bigger plants.

Herbivorous zooplankton, for example copepods, eat phytoplankton. Some of the food they consume will pass through the gut and be lost as faeces. Of the remainder, that which is **assimilated**, some will be used for **respiration** and some will be used for **secondary production**, i.e. for copepod growth and reproduction. These relationships are summarised in Fig 1.8. Secondary production represents food which is potentially available to carnivorous zooplankton, such as arrow worms and herring. Again, some of the food that these organisms consume will be lost as faeces while some will be used for respiration and production. In the case of fish it is this production that is available to humans as food.

At each stage in the food chain, some of the chemical energy in the food consumed is lost as heat produced during respiration. At the same time, some of the organic material formed by each trophic level sinks out of the surface layers of the sea. Dead and partially eaten phytoplankton and

THE NATURAL HISTORY OF THE SEA

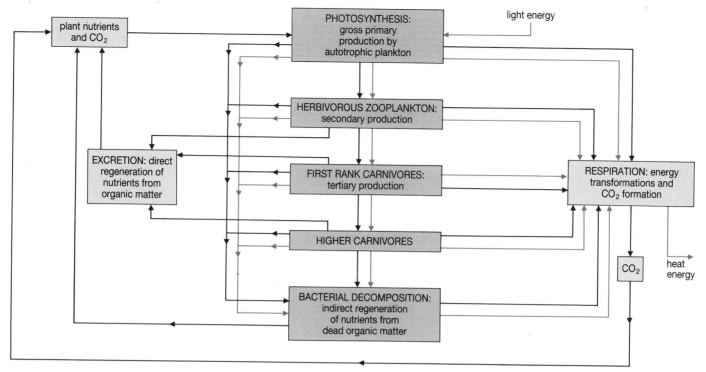

Fig 1.8 Energy flow and nutrient cycles in the sea.

zooplankton, zooplankton faecal pellets and the moulted exoskeletons of crustaceans like copepods (collectively called **detritus**) provide a source of food for both animals and bacteria. The deeper down in the sea the more important this supply of food becomes. So to complete our picture of the pelagic food web shown in Fig 1.7 we need to include not only grazing food chains but decomposer food chains as well.

To summarise: the energy of sunlight is converted into chemical energy, during the process of photosynthesis, by autotrophic plankton. The organic molecules so produced provide the food for herbivorous zooplankton, which in turn provide food for carnivorous zooplankton and fish, the **grazing food chain**. In addition, a large proportion of primary production, which includes both cells and organic substances, like glycolic acid, which leak out of autotrophic cells as they photosynthesise, is consumed by bacteria. These bacteria in turn form the food of protozoa, forming the so-called **decomposer food chain** (Fig 1.9). As more marine food chains are studied it is beginning to emerge that the decomposer food chain is actually of greater importance than the grazing food chain in pelagic food webs. Notice from Fig 1.9 that part of the microbial production is returned to the herbivore food chain via the so-called 'microbial loop'.

Benthic food chains

While some of the organic material, aptly termed **marine snow**, falling from the surface layers of the sea may be consumed by detritivores, or broken down by bacteria, some will eventually reach the sea bottom. This provides the food source for benthic food webs.

In the benthos (i.e. the layer of benthic organisms) some of the detritus will be consumed directly by large animals such as shellfish. However, much of the material will be metabolised initially by bacteria and protozoa. These microorganisms then form the basis of benthic food chains. At the top of these food chains are the **demersal** (bottom living) fish, such as plaice and cod, which feed on the larger worms, crustaceans and molluscs.

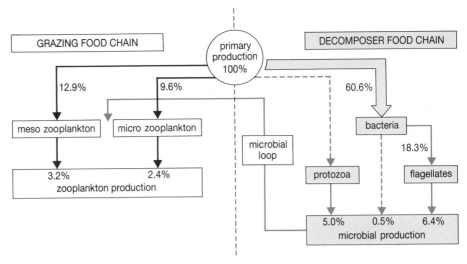

Fig 1.9 A carbon budget for the English Channel in the summer. Note that at this particular location over 60% of the total carbon fixed by the autotrophic plankton went to decomposers. An unknown amount of microbial production passes back to the grazing food chain via a microbial loop.

In some areas, plaice also have a habit of biting off the protruding siphons of buried bivalve molluscs, leaving their owners otherwise unharmed and apparently able to regenerate their missing pieces.

Nutrient cycles

Pelagic and benthic bacteria play an essential role in breaking down detritus and hence in regenerating the nutrients needed by the autotrophic plankton for growth. While some of this regeneration will take place in the sunlit surface waters the continual loss of detritus as it sinks means that there will be a gradual loss of nutrients from the sea surface. Plants can photosynthesise for a short period when supplied with only carbon dioxide and water, but if they are denied essential nutrients, photosynthesis will eventually stop. The logical consequence of this is that primary production would eventually cease and all the organisms in the sea would die. Obviously, this does not happen so there must be some process which returns nutrients from the bottom of the sea to the surface waters.

QUESTIONS	

1.5 (a) Using Fig 1.7 give two examples of (i) primary producers (ii) secondary producers (iii) secondary consumers.
(b) How many trophic levels does the herring feed at?

1.6 What substances available in sea water will primary producers need to make
(a) carbohydrates
(b) fats
(c) proteins
(d) nucleic acids?

1.4 THE MOVEMENT OF WATER IN THE SEA

Consider a piece of detritus which has reached the bottom of the sea. It will be colonised rapidly and broken down by microorganisms, releasing essential nutrients like nitrates and phosphates. To be incorporated into new tissue, these nutrients must be returned to the sea surface where they can be used by the autotrophic plankton. How can this movement be achieved? Well, the nutrients could diffuse back to the surface, but as this process is very slow it is unimportant as a method of returning nutrients to the surface. What is needed is a bulk transport system. Just as the blood in your body carries oxygen from your lungs to your toes, so the movement of sea water carries nutrients from the deeper layers of the sea back to the surface. So we need to find out about the processes which move sea water.

Waves and turbulence

As the wind blows over the surface of the sea, some of its kinetic energy is transferred to the water by friction, leading to the generation of waves (Fig 1.10). These surface waves will cause turbulence in the underlying water, leading to vertical mixing of the water column. However, this mixing will not occur to any great depth since the motion produced by the waves declines rapidly with depth (see Fig 1.10). The greater the velocity of the wind the larger the waves produced and so the greater the depth to which turbulent mixing will occur. Such mixing, which may occur to a depth of 200 m when the wind is very strong, for example during winter gales, will return nutrients from deeper water to the surface.

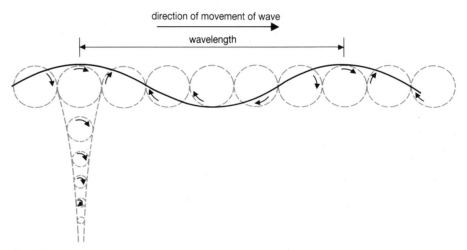

Fig 1.10 A profile of a wave. The circles and arrows show the direction of movement of water in different parts of the wave. Note how the water movement decreases rapidly with depth.

Surface currents

In addition to creating waves the wind also produces surface currents (Fig 1.11) when it blows in the same direction over long periods of time. Note the following points from Fig 1.11:

- The circular flow of water around the major ocean basins, e.g. the North Atlantic. The water flows clockwise in the northern hemisphere and counter-clockwise in the south. These circular systems are called **gyres**.
- The equatorial currents and counter-currents, most noticeable in the Pacific.
- The Antarctic circumpolar current.

The movement of water by these surface currents affects local climatic conditions and thus the marine organisms found. For example, warm water carried across the Atlantic by the North Atlantic drift keeps the coast of Ireland and Norway ice free while coasts at the same latitude on the other side of the Atlantic, e.g. Greenland and Labrador, are ice bound.

Upwellings

Surface currents not only move water horizontally but may cause it to move vertically, so producing an **upwelling**. This is a process whereby nutrient-rich, cool water rises to the surface from the depths. Such water provides an ideal medium for the growth of phytoplankton, principally diatoms, which increase greatly in abundance as a result. Upwellings are of widespread occurrence but to be of importance in returning nutrients to surface waters, and so enhancing planktonic primary production, they must occur for long periods of time. Such long-lived upwellings are characteristically found in three places.

1. The slow moving currents found on the eastern side of continents (Eastern boundary currents), e.g. the Benguela, Canary and Peruvian currents (see Fig 1.11).

2. The equatorial currents.

3. The Antarctic circumpolar current.

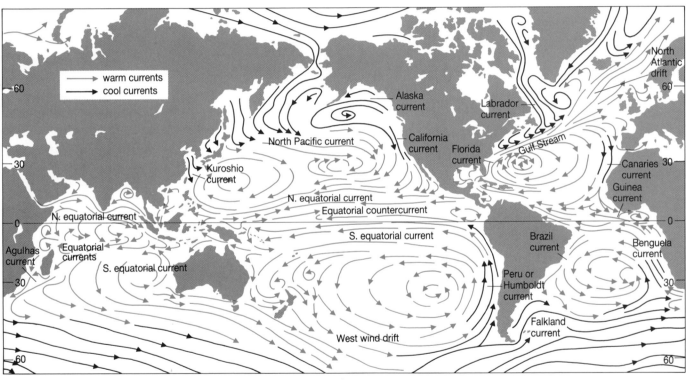

Fig 1.11 The surface currents of the world's oceans. Warm currents are shown in red, cold ones in black. Note the large gyres (circular movements of water) in the Pacific and Atlantic.

Eastern boundary current upwelling

An upwelling system typical of these currents is shown in Fig 1.12. Such upwellings develop when the wind blows over the sea parallel to the shore and towards the equator. As a result, surface water is moved away from the coast by a process called **Ekman transport**, with the surface water being replaced by water upwelling from below, usually from depths of 50–300 m. In the Benguela current, such an upwelling zone may be 65–130 km across and 1600 km long. The water which upwells, although coming from only a relatively shallow depth, is rich in inorganic nutrients like phosphates and nitrates.

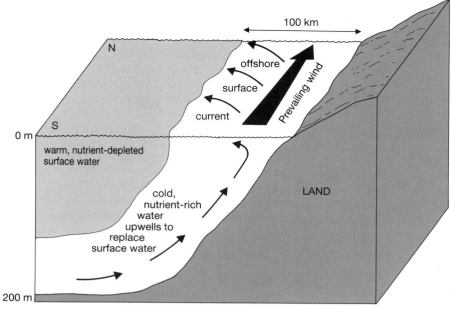

Fig 1.12 An upwelling in an eastern boundary current. Note that water is being moved offshore by a wind blowing along the coast, a process called Ekman transport.

Equatorial current upwelling

Fig 1.13 shows the equatorial current system in more detail. Note that there are three currents.

1. The south equatorial current (SEC) straddling the equator.

2. The north equatorial current (NEC).

3. The Equatorial Counter Current (ECC) sandwiched between them.

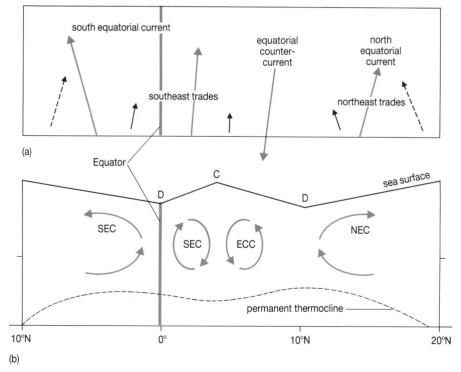

Fig 1.13 Upwelling in equatorial currents. The currents are shown in red and the wind direction by the black arrows. **(a)** View from above. **(b)** Surface currents in cross-section. The slope of the sea surface is greatly exaggerated. D = divergence where upwelling occurs; C = convergence.

In this system there are two areas of upwelling and one of downwelling. The flow of water in the south equatorial current diverges as the water flows westwards, producing an area of divergence (D) in the centre of the current. Deep water will upwell into this zone of divergence. On the northern side of the south equatorial current and the southern side of the equatorial counter-current there is a zone of convergence (C) where surface water downwells. Finally, there is another area of upwelling on the northern side of the equatorial counter-current and the southern side of the north equatorial current. Again, the upwelled water will be rich in nutrients.

Antarctic current upwelling

Contact with the ice covering Antarctica cools the sea water surrounding the continent. The cooler sea water is more dense and so it sinks. Water upwells to replace the sinking water, again bringing nutrients with it.

Bottom currents

The water which sinks along the coast of Antarctica flows northwards as it descends over the continental shelf and slope to the abyssal plain. This Antarctic bottom current flows northwards at a rate of some 20 million m^3 of water per second, reaching well into the northern hemisphere. A similar but smaller bottom current flows south from the Arctic. While these currents are irrelevant to planktonic organisms they do carry enormous amounts of dissolved oxygen, so ensuring that there is sufficient oxygen in the deep oceans to meet the respiratory needs of deep-sea organisms.

QUESTIONS

1.7 (a) Explain the following observations.
   (i) Most of the whales harvested for their oil and meat prior to the ban on whaling were killed in Antarctic waters.
   (ii) Other krill-feeding Antarctic animals, like penguins and Weddel seals, have increased in number since the decline of baleen whales through over-hunting.
   (b) What effects do you think harvesting krill would have on the recovery of whale stocks and the numbers of penguins?

1.8 What would the effect of a current convergence have on the growth of phytoplankton? Explain your answer.

1.5 THERMOCLINES

The temperature of the water at different depths in the sea can be assessed using electronic thermometers. This data can then be used to produce a graph which shows how temperature changes with depth, a **temperature profile**.

Fig 1.14 shows temperature profiles for different latitudes. In the polar oceans the temperature is practically the same from top to bottom although cold, low salinity water near the surface disturbs the otherwise vertical profile. Elsewhere the intense heating caused by sunlight results in the formation of a light, warmer layer of water floating on top of a much larger mass of cold, dense water. Where these two bodies of water meet there is a zone of rapid change in water temperature, called the **thermocline** (Fig 1.14(b)). This thermocline is a permanent feature of both tropical and temperate waters, although it is less well developed, and at a deeper depth, in temperate compared with tropical oceans.

When the surface water of temperate seas is warmed in the summer a further seasonal thermocline can develop much closer to the surface

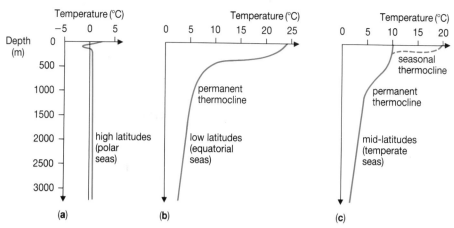

Fig 1.14 Temperature profiles for different latitudes in the open ocean. Note that the almost vertical profile for polar oceans is slightly disturbed near the surface by cold water from melting ice.

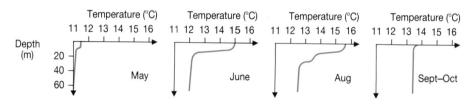

Fig 1.15 The development and destruction of a seasonal thermocline in the English Channel.

(Fig 1.14(c)). This is a much stronger feature than the permanent thermocline in temperate waters. Fig 1.15 shows the development of such a thermocline in the English Channel. As we will see in the next chapter this thermocline plays an important role in determining seasonal changes in marine primary production in temperate seas. For the moment you should note the following factors.

- **Thermal stratification**, and the development of a thermocline, represents a physical barrier to the mixing of the water column.
- There is little or no exchange of nutrients through an established thermocline.
- Effectively, once nutrients locked up in detritus falling from the euphotic zone have passed below the thermocline they are lost to the surface until the thermocline is destroyed by vertical mixing of the water column.
- The stronger, i.e. the steeper, the thermocline the more effective this nutrient prison becomes.
- Thermoclines are an important factor in the control of marine primary production. This is discussed in the next chapter.

QUESTION	1.9 (a) The southern North Sea has an average depth of 60 m. Would you expect to find a permanent thermocline in this shallow temperate sea?
	(b) How will upwellings and turbulent mixing impede the development of a thermocline?
	(c) Is the thermocline found at the bottom of the euphotic zone? Explain your answer.

SUMMARY ASSIGNMENT

1. Make a copy of Fig 1.1 and annotate it to show how light intensity, temperature, pressure, oxygen concentration and nutrient levels change with depth.

2. What is the photic (euphotic zone)? What factors do you think will affect the depth of the photic zone? Explain why the photic zone will be deeper in the centre of a subtropical oceanic gyre than in coastal waters subject to the effects of waves, tide and river run-off.

3. Distinguish between the following terms: (i) autotroph, heterotroph; (ii) photosynthesis, chemosynthesis; (iii) phytoplankton, zooplankton; (iv) pelagic organisms, benthic organisms; (v) holoplankton, meroplankton; (vi) oceanic and neritic waters;

4. Explain how plankton can be classified on the basis of (i) size (ii) taxonomy (iii) ecological roles. What is the advantage of an ecological classification?

5. What is a primary producer? Give a brief account of three primary producers found in the sea. In what form may primary production be consumed by heterotrophs?

6. Discuss the relative importance of grazing and decomposer food chains in pelagic and benthic food webs. How do benthic food webs harvest the energy of sunlight?

7. How are inorganic nutrients like nitrates and phosphates regenerated from organic molecules in the sea? Where does this regeneration mainly occur and how are the nutrients returned from here to the surface water? Why is it important that this should occur?

8. What is a thermocline? Distinguish between permanent and seasonal thermoclines. How are thermoclines established and how do they vary with latitude? Why are thermoclines of importance to autotrophic plankton?

Chapter 2

MARINE PRODUCTIVITY

Imagine you are the captain of a fishing vessel. How do you decide where to fish? Experience will probably tell you that place A will produce the best catch at this time of the year. But why is place A better than place B or place C? One possible reason is that there is more food available for fish to eat at place A, and more food means more fish. So a good place to start when looking at fisheries ecology is the processes that produce fish food.

LEARNING OBJECTIVES

After completing the work in this chapter you will be able to:

1. define the following terms: gross primary production, net primary production, primary productivity, standing crop, secondary productivity, compensation light intensity, compensation depth, critical depth;

2. describe and account for patterns in global marine primary productivity;

3. explain how the effects of light intensity, nutrient availability and turbulence interact to control marine primary productivity;

4. describe and explain seasonal variations in primary productivity;

5. calculate and explain the significance of the ecological efficiency of marine food chains.

2.1 PRIMARY PRODUCTION IN THE SEA

So far we have used the term primary production rather loosely. We now need to define it more carefully.

Gross and net primary production

During photosynthesis the energy contained in photons of light is changed into chemical energy by converting (**fixing**) carbon dioxide into organic molecules like glucose. This process is summarised in Fig 2.1. Note that to produce organic molecules like proteins and nucleic acids a supply of nutrients, nitrates, phosphates and so on, is required in addition to carbon dioxide and water.

Only a small amount of the solar energy falling on the sea, about 0.1–0.5%, is actually converted into the chemical energy of newly synthesised organic molecules by autotrophs. The total amount of energy fixed by photosynthesis is called the **gross primary production (GPP)**. Part of this energy is used by the plant for **respiration (R)** and is lost as heat. There is therefore an equilibrium, **the compensation point**, where the rate of production of organic molecules by photosynthesis is balanced by the rate of their destruction by respiration. If autotrophs always existed at the compensation point there would be no production of food for animals to eat, since all the carbohydrate formed by photosynthesis would be consumed by the plants' own respiration. The fact that there is food for herbivores means that autotrophs must live at least part of their lives

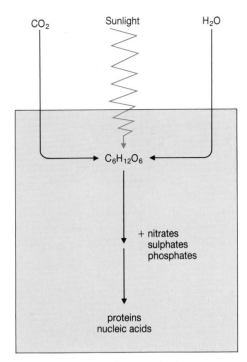

Fig 2.1 A summary of photosynthesis.

above the compensation point. We can therefore define the **net primary production (NPP)** as follows:

Net primary production = energy fixed in photosynthesis – energy lost in respiration
(NPP) (GPP) (R)

that is:

$$NPP = GPP - R$$

Primary production and standing crop

We now need to distinguish between the rate of primary production, the **primary productivity**, and the **standing crop**, the total biomass of a particular organism in a given area or volume of water at a particular time. You can begin to appreciate the difference by thinking about a lawn. During the summer you mow the grass every week so that the standing crop, the amount of grass on the lawn, is always quite small. But the pile of grass clippings, which represents the net primary production of the lawn through the summer months, will be enormous. Now think of a population of autotrophic plankton (equivalent to the lawn) being grazed by zooplankton (equivalent to the lawn mower). A small standing crop of autotrophs can support a large population of zooplankton provided the **productivity** (the rate of production of new biomass) of the autotrophs is high.

In ecology, primary production and standing crop are often measured in units of energy, e.g. kilojoules per square meter (kJ m^{-2}). However, in marine ecology, these values are often measured in terms of grams of carbon per square metre (gC m^{-2}). This figure refers to all the plant carbon in a water column of area one meter square, extending from the water surface to the sea bottom. The rate of primary production, the primary productivity, is then measured as gC m^{-2} day^{-1} or gC m^{-2} yr^{-1}. For large values, tonnes (1 tonne = 1000 kg) are used instead of grams (tC m^{-2} yr^{-1}).

The arguments advanced for primary production are equally valid for **secondary production**, i.e. production by heterotrophs. What is important is how fast new herbivore tissue is being formed, not how much is present at any one time. It is this figure which determines how much food is being made available for consumption at higher trophic levels.

Global marine primary productivity

Table 2.1 gives data on the biomass and productivity of marine autotrophs. Overall, marine global primary productivity has been estimated at 25 × 10^9 tC yr^{-1} of which 4 × 10^9 t is thought to be fixed annually on the continental shelves and upwelling areas, although here again there is

Table 2.1 A marine algal biomass and productivity.

Province	Area (10^6 km^2)	Total biomass (10^6 tC*)	Total net primary productivity (10^6 tC yr^{-1})
open ocean	332.0	455	18 900
upwelling zones	0.4	3.6	90
continental shelf	26.6	123	4 000
algal beds and reefs estuaries (excluding marsh)	0.6	545	700
total	361.0	1773	25 000

*tC = tonnes of carbon – a measure of productivity

Table 2.2 Average daily rate of phytoplankton production in continental shelf waters of various oceans, and in various areas of upwelling.

Area	Net Primary Productivity (gC m^{-2} day^{-1})
continental shelves*	
Atlantic Ocean	0.41
Pacific Ocean	0.52
Indian Ocean	0.71
Antarctic	0.89
upwelling areas	
Peru, Canary, Benguela	0.3–1.0
California, NW Australia	0.3

* Upwelling areas included.

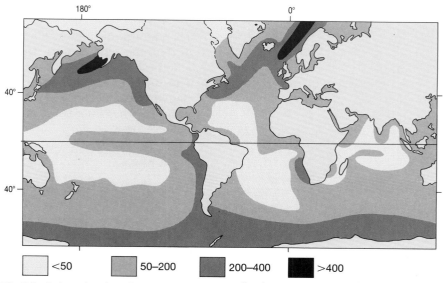

| | <50 | | 50–200 | | 200–400 | | >400 |

Fig 2.2 Estimated marine primary productivity in gC m^{-2} yr^{-1}.

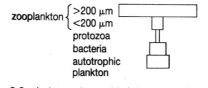

Fig 2.3 An inverted pyramid of biomass from the English Channel. Note the width of the bars is proportional to the biomass at each trophic level.

considerable variation (Table 2.2). Fig 2.2 gives a visual impression of marine productivity. The main features for you to note from Fig 2.2 are:

• the considerable variation in marine primary productivity;
• coastal areas are more productive than mid-oceanic regions;
• the high productivity of the polar seas;
• the zone of higher productivity lying around the equator in the Pacific Ocean;
• the high productivity in areas of upwelling.

To explain these patterns we now need to consider the factors which control primary production in the sea.

QUESTION

2.1 (a) Explain why the total standing crop of marine primary producers is only 1773 × 10^6 tC while the total NPP is 25 000 × 10^6 tC yr^{-1} (see Table 2.1).

(b) From the data in Table 2.1 calculate which areas have the greatest biomass and productivity per unit area. Explain, as far as you can, the results you obtain.

(c) Inverted pyramids of biomass (Fig 2.3) are common in planktonic food chains. Account for this observation.

2.2 FACTORS LIMITING MARINE PRIMARY PRODUCTIVITY – LIGHT

Let us quickly summarise some basic information. The sun represents the only external energy source for the Earth and the only way this energy can be converted into chemical energy is by photosynthesis. In the sea the ultimate food source for all heterotrophs is the primary producers, plants and some bacteria, which use sunlight, carbon dioxide, water and inorganic and organic nutrients (e.g. vitamin B_{12}) to produce organic molecules. We can discount chemosynthesis since it only contributes about 5% of marine primary productivity, although it may be important in some habitats.

Light penetration

Water absorbs light readily (Fig 2.4). More than half of the solar radiation is absorbed in the first metre of water including almost all of the longer wavelength infrared radiation. Indeed, about one-quarter of the energy of sunlight is absorbed within 3 mm of the surface! So even in clear water only about 5–10% of the sunlight may be present at a depth of 20 m.

The penetration of light into the water column will be affected by two factors.

1. The amount of light available, which in turn is determined by the time of day, the time of year, the latitude and the weather, particularly the amount of cloud.
2. The turbidity of the water, because the more plankton and particles (together called the **seston**) the water contains the less light will penetrate. This has interesting consequences for NPP in coastal waters as we will see later.

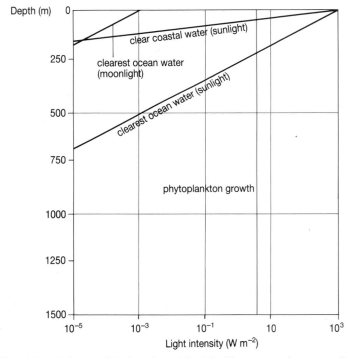

Fig 2.4 The relationship between light intensity and depth in different types of sea water. The red vertical line shows the minimum light intensity needed for photosynthesis. Note that light intensity is shown on a logarithmic scale.

Compensation depth

Fig 2.5 shows the relationship between light intensity and the rate of photosynthesis. Since a certain amount of the gross primary production of autotrophic plankton will be used up in respiration there must be some

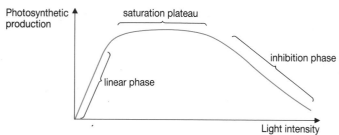

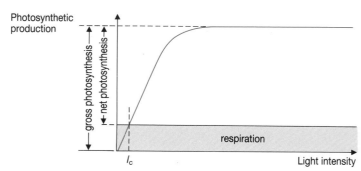

Fig 2.5 The relationship between light intensity and photosynthesis.

Fig 2.6 Derivation of the compensation light intensity (I_c) and the relationship between gross and net primary production.

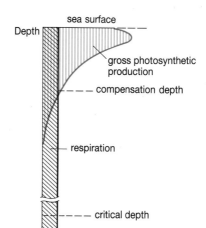

Fig 2.7 The relationship between gross primary production, respiration and depth. The compensation depth is the depth at which the oxygen consumed by respiration equals that released by photosynthesis over a 24-hour period. The critical depth is the depth at which the area under the photosynthesis curve, i.e. the total amount of photosynthesis in the sea above, equals that under the respiration curve.

light intensity, called the **compensation light intensity (I_c)**, where the rate of photosynthesis is exactly balanced by the rate of respiration (Fig 2.6). If the light intensity exceeds I_c then organic compounds will be accumulated by the autotrophic cells and they will grow and divide. The amount of growth and reproduction is the NPP: the food available to herbivores and decomposers.

Since depth is a major factor affecting light intensity in the sea, we might expect there to be some relationship between depth and compensation light intensity. This is shown in Fig 2.7. (This is an important diagram so it needs careful study.) Note that the amount of respiration is assumed to be the same at all depths and is usually considered to use 10% of GPP achieved at saturation light intensity (see Fig 2.5). The amount of photosynthesis, i.e. GPP, follows the same pattern as before. If you follow the 'photosynthesis' curve (in red) downwards from the surface you will eventually reach a depth where the rate of photosynthesis equals the rate of respiration. This is the **compensation depth**. (Note that the water lying above the compensation depth represents the **euphotic zone**.) As the compensation depth will change with the time of day, it is usually defined as the depth at which the amount of carbon fixed by photosynthesis over a period of 24 hours is equal to the amount of carbon used in respiration during that 24 hours. In practice, rules of thumb are used to establish the compensation depth, for example the depth to which 1% of the surface light penetrates. Remember, while we are talking about a depth the real variable involved is light intensity.

QUESTIONS

2.2 (a) From Fig 2.4 could photosynthesis proceed (i) on a moonlit night (ii) below 100 m in sunlit coastal waters (iii) below 200 m in clear sunlit ocean water?

(b) What will the predominant wavelength of light be at a depth of 20 m?

(c) What is the adaptive advantage of accessory pigments to autotrophic plankton?

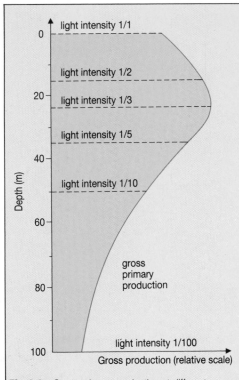

Fig 2.8 Gross primary production at different depths with light intensities expressed as fractions of the intensity at the sea surface.

2.3 (a) Using Fig 2.5 describe the relationship between light intensity and photosynthesis.
(b) From your knowledge of the light and dark reactions of photosynthesis suggest the factors which limit the rate of photosynthesis in the (i) linear phase (ii) saturation phase (iii) inhibition phase.

2.4 The relationship between depth and the rate of photosynthesis at a tropical site on a clear day is shown in Fig 2.8.
(a) At what depth does the maximum rate of photosynthesis occur?
(b) Why does the maximum rate of photosynthesis not occur at the surface?
(c) Why does the rate of photosynthesis decrease below about 23 m?

2.5 (a) What is the relationship between the compensation depth and the compensation light intensity?
(b) How will the turbidity of the water affect the compensation depth?
(c) How will the compensation depth change with time of year in (i) tropical seas (ii) temperate seas (e.g. those off the coast of Britain) (iii) polar seas?
(d) (i) Contrast the fate of an autotrophic cell trapped above the compensation depth with one trapped below it. (ii) Does this explain why so many planktonic autotrophs have structures which inhibit their rate of sinking? Explain your answer.
(e) Does photosynthesis occur below the compensation depth?

2.3 FACTORS LIMITING MARINE PRIMARY PRODUCTIVITY – NUTRIENTS

If light intensity was the only factor affecting primary productivity then we might expect the greatest rates of primary production to occur in tropical seas with polar seas being much less productive. However, if you look at Fig 2.2 you can see that this is clearly not the case. With the exception of areas of upwelling, productivity in tropical seas, particularly in the centre of oceanic gyres, is very low. In fact, some of the highest productivities occur on the continental shelves of temperate seas, and in polar waters where the light intensity is very low for a substantial part of the year. Clearly, then, some other factor, in addition to light intensity, is controlling marine primary production. The most likely candidate is the availability of nutrients.

The concept of limiting nutrients

The biologically important elements in sea water can be grouped into three categories:

1. Bio-unlimited, e.g. Na, Cl, K.
2. Bio-intermediate, e.g. C, Ca, Ba.
3. Bio-limiting e.g. P, N, Si and possibly trace elements such as Cu, Ni.

In addition, some autotrophic plankton require organic nutrients like vitamin B_{12}.

Elements which are bio-limiting can be defined as those whose availability limits primary production. Although there is much controversy among marine biologists about the extent to which nutrient availability actually limits primary production, the general consensus is that three elements are of particular significance in sea water.

1. Nitrogen – primarily as nitrate, but also ammonia and nitrite.
2. Phosphorous – mainly in the form of phosphate although the importance of organic phosphates is increasingly recognised.

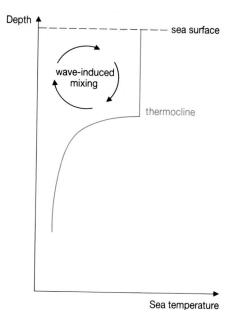

Depth

sea surface

wave-induced mixing

thermocline

Sea temperature

Fig 2.9 Once a thermocline has become established the surface water does not mix with the water below the thermocline.

3. Silica – certain algae, for example diatoms and silicoflagellates (Chrysophyceae), form cell walls which contain silica. Such cells have an absolute requirement for silica to grow.

Nutrient depletion

Fig 2.9 shows a column of water where a strong thermocline has become established. Note that vertical mixing is only occurring in the layer of water above the thermocline. This layer is well illuminated but as photosynthesis and primary production proceed, more and more of the nutrients in the euphotic zone will become incorporated into organic molecules like proteins and DNA. Some of these nutrients will be recycled rapidly through herbivore grazing and excretion, as well as by decomposition by microorganisms. However, many of the nutrients, locked up in detritus, will fall through the thermocline and be lost to primary producers in the euphotic zone. As a result of this process the surface waters will become gradually depleted of nutrients.

Eventually, there will come a point when the concentration of nutrients is so low that the autotrophic cells will not be able to take up the ions from the sea water because their active transport mechanisms can no longer cope with the enormous difference in concentration of ions between cytoplasm and sea water. At this point the cells will have to start using nutrients that they may have stored when nutrient concentrations were higher. However, once this store is depleted the rate of primary production will fall dramatically.

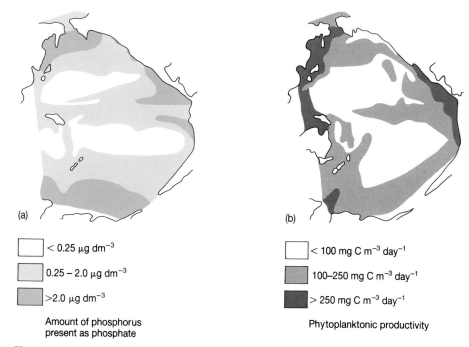

(a)

$< 0.25 \ \mu g \ dm^{-3}$

$0.25 - 2.0 \ \mu g \ dm^{-3}$

$>2.0 \ \mu g \ dm^{-3}$

Amount of phosphorus present as phosphate

(b)

$< 100 \ mg \ C \ m^{-3} \ day^{-1}$

$100{-}250 \ mg \ C \ m^{-3} \ day^{-1}$

$> 250 \ mg \ C \ m^{-3} \ day^{-1}$

Phytoplanktonic productivity

Fig 2.10 **(a)** The distribution of phosphate (measured at a depth of 100 m) and **(b)** Primary productivity in the Pacific Ocean.

QUESTIONS

2.6 Fig. 2.10 shows the distribution of phosphate and primary production in the Pacific Ocean. This sort of correlational evidence has been used to suggest that nutrient availability limits marine primary productivity.
(a) What evidence is there from Fig 2.10 that phosphate limits primary productivity?

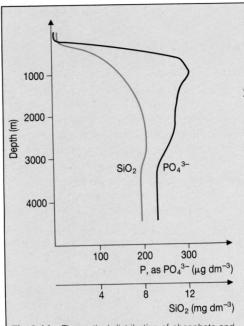

Fig 2.11 The vertical distribution of phosphate and silicate at 24°22' N, 145°33' W.

(b) Is this evidence proof that phosphate limits primary productivity?

(c) What additional evidence would be needed to provide such proof?

2.7 Fig 2.11 shows the vertical distribution of phosphate and silicate in a water column.

   (a) What is the location of this site in relation to the major ocean gyres?

   (b) What is the concentration of phosphate and silicate at (i) the surface (ii) a depth of 1000 m?

   (c) Would you expect nitrate to follow the same pattern?

   (d) Account for the general shape of these curves.

   (e) Predict what would happen to the shape of the curves if the water column was subjected to vertical mixing.

   (f) Explain why primary productivity is so high in tropical upwelling zones.

   (g) Explain why the primary productivity of tropical seas in the absence of upwelling is so low.

2.4 FACTORS LIMITING MARINE PRIMARY PRODUCTIVITY – TURBULENCE

We now need to pull a few strands together. Let us first summarise what we already know about the control of primary production in the sea.

- All photoautotrophic production, which comprises 95% of all marine primary production, necessarily occurs in the illuminated water of the euphotic zone.

- The great bulk of bacterial decomposition occurs in the deeper waters of the sea.

- When a body of water is heated from above and when there is little vertical mixing of the water column, for example when there is little wind, the water column becomes stratified to form an upper, warm, mixed layer overlying a cooler, lower layer, with a layer of rapid temperature change, the thermocline, lying between them. The thermocline acts as a physical barrier to water circulation, thereby inhibiting the supply of nutrients from deeper to surface waters.

- If the thermal stratification of the water column persists for long enough the nutrients become depleted in the euphotic zone, so reducing primary productivity.

- This effect is particularly pronounced in tropical seas where, in the absence of upwelling, there is a permanent, well-developed thermocline close to the surface. The seasonal thermocline which develops during the summer in temperate seas has a similar effect.

Critical depth

We now need to define another term which will be vital to our understanding of the relationship between primary production and the depth of mixing of the water column. Look back at Fig 2.7. There must be some depth in that column of water where the amount of GPP is exactly equalled by the amount of respiration. This is the **critical depth** which is defined as the depth at which the amount of GPP in the water column above is exactly equal to the amount of respiration.

ANALYSIS

The effects of turbulence

This exercise will help you to practise analytical thinking skills.

Look at Fig 2.12. Fig 2.12 (a) shows the temperature profile of a column of water with a shallow, mixed layer. Fig 2.12(b) shows the distribution of net primary production and Fig 2.12(c) shows the nitrate concentration in the same column of water.

(a) At what depths are (i) the thermocline and (ii) the compensation depth in Fig 2.12(a). Explain why the compensation depth is below the thermocline.

(b) Account for the changes in nitrate concentration with depth.

(c) Consider a photoautotrophic cell above the thermocline. Will such a cell be (i) taken below the critical depth by vertical mixing (ii) be light limited (iii) be nutrient limited?

Now imagine that the wind velocity increases. The nitrate distribution in the column of water is now as shown in Fig 2.12(d).

(d) Explain why (i) the depth of the mixed layer has increased
(ii) the amount of nitrate in the euphotic zone has increased.
(iii) What do you think will happen to the thermocline under these circumstances? (iv) If the photoautotrophs in the euphotic zone had been nutrient limited what effect will the increased depth of mixing have on NPP?

Now look at Fig 2.12(e). The mixed layer now extends below the critical depth. Think about an individual algal cell. It is being continually circulated by turbulence and, under these conditions, spends much of its time below the critical depth. So the average light intensity experienced by the cell is less than the compensation light intensity. This means that the cell now uses up more organic molecules in respiration than it is making in photosynthesis.

(e) Predict what will happen to NPP under the circumstances shown in Fig 2.12(e).

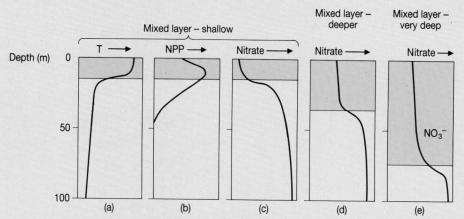

Fig 2.12 The relationship between increased mixed layer depth and net primary production NPP. T = temperature.

Turbulent vertical mixing of the water column has two opposite effects. On the one hand, it causes upwelling, bringing nutrients into the euphotic zone and so increasing primary productivity. On the other hand, when the mixed layer exceeds the critical depth it results in a decrease in NPP because the rate of photosynthesis becomes light limited. It is the balance between these two effects of turbulence which is the most important factor controlling marine primary productivity.

In tropical seas where, in the absence of upwelling, there is permanent stratification with a strongly developed thermocline and little vertical mixing, primary production is nutrient limited. Since the availability of nutrients in the euphotic zone is low, primary production in these waters is also low.

The same situation applies in many temperate waters in the summer as a result of the formation of a strong seasonal thermocline. However, the same temperate waters will be cooled and subjected to gales during the autumn and winter, with the result that the water is mixed to a depth of up to 200 m, so destroying the thermocline. Nutrient availability in the euphotic zone is now high, but as the photoautotrophic cells are being circulated below the critical depth, primary production is low because it is now light limited.

Neritic waters

In coastal, or neritic waters, other forces, in addition to wind, act to create vertical mixing. These include the effects of tidal currents and river run-off. The net effect is that there is almost continuous upwelling of nutrient-rich water. Under these circumstances, phytoplankton biomass may be very high, the water highly turbid, and the extinction of light with depth very rapid. Consequently, the compensation depth will be very close to the surface.

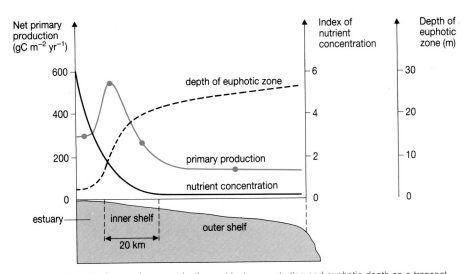

Fig 2.13 Variation in net primary production, nutrient concentration and euphotic depth on a transect from the coast of Georgia, USA, to the edge of the continental shelf.

2.5 SEASONAL VARIATION IN MARINE PRIMARY PRODUCTIVITY

We now have an explanation of the pattern of primary productivity shown in Fig 2.2. Productivity tends to be high in waters where nutrients are readily available, provided sufficient light is available. But there must be a further factor affecting this pattern because primary productivity is very high in polar waters where light availability is very low for six months of the year. This suggests that there must be tremendous seasonal variation in primary productivity as well as spatial variation. We now need to look at this aspect of marine primary productivity.

Fig 2.14 shows the seasonal variation in standing crop of autotrophic plankton and zooplankton in a northern temperate ocean. The important thing to realise here is that this pattern is largely determined by exactly the same factors which determine spatial variation in biomass – the availability of light and the availability of nutrients. These will be affected by the degree of turbulent mixing of the water column, the presence or absence of a thermocline, and the turbidity of the water. We will go through Fig 2.14 season by season and so piece together the underlying story.

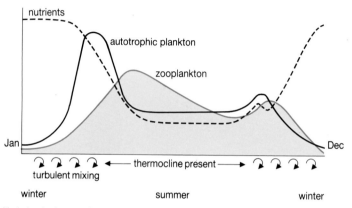

Fig 2.14 Variation in the standing crop of autotrophic plankton and zooplankton in the euphotic zone of a northern temperate ocean.

Winter

Winter gales and cooling break down the seasonal thermocline as the warm surface waters become mixed with the underlying cooler waters. As a result the surface layers become rich in nutrients. However, the low light intensity means that the critical depth is close to the surface. As a result the depth of the mixed layer exceeds the critical depth and so photosynthesis is light limited. Productivity and standing crop of both autotrophic plankton and zooplankton are low.

Spring

The concentration of nutrients in the surface layers is still high and since the light intensity is increasing the critical depth increases. At the same time the turbulence is not so intense as the winter gales subside and the depth of the mixed layer does not exceed the critical depth. Under these circumstances, diatoms and other autotrophs start to grow and divide rapidly, leading to the increase in autotrophic plankton, or spring bloom.

Summer

During the spring bloom, nutrients are used up rapidly, but these are replaced from deeper water by mixing. However, as spring progresses the seasonal thermocline begins to develop, both as a result of intense heating and the further reduction in the amount of wind. So by late Spring (often

May) nutrient availability above the thermocline is very low. Under these circumstances, large diatoms, like *Skeletonema*, do not seem able to take up nutrients although smaller autotrophic plankters and cyanobacteria can and may still flourish. However, since these are smaller cells the overall standing crop of autotrophic plankton falls. Note here that we are talking about standing crop not productivity.

The limiting factor involved in ending the spring bloom seems, then, to be nutrient depletion, although two other factors are also likely to be important.

1. Increased grazing by zooplankton, such as copepods: Remember that primary producers can sustain high rates of grazing even when their standing crop is low, provided productivity is high. If, however, productivity is low, because of nutrient depletion, then grazing will soon reduce the standing crop of primary producers. This is equivalent to somebody who has been living off the interest earned on a sum of money they have invested suddenly finding the interest rate (equivalent to productivity) has been reduced to virtually zero. They then have to live off the capital (equivalent to the standing crop) which soon disappears.
2. Turbidity: The great increase in diatom numbers during the spring bloom will tend to increase the turbidity of the water and so reduce the compensation depth. If this shading becomes very severe it may result in a reduction in NPP and so a fall in biomass.

Autumn

As the surface waters cool, thermal stratification decreases and the thermocline begins to break down. This process is accelerated by autumn gales. As a result, deeper, cooler, nutrient-rich water upwells, while there is still sufficient light to sustain high rates of photosynthesis. Under these conditions, diatoms begin to grow and divide again, leading to a further increase in standing crop. The autumn increase in the temperature of seas is not as universal as the spring increase and is of a more limited extent.

QUESTIONS

2.10 **(a)** What is the fundamental difference in the mechanism underlying the origin of the spring and autumn diatom increase?
 (b) What might cause the autumn decline?

2.11 Fig 2.15 shows two other patterns of temporal variation in plankton biomass.
 (a) Identify the latitudes which these graphs relate to.
 (b) Describe and explain the patterns you can see.

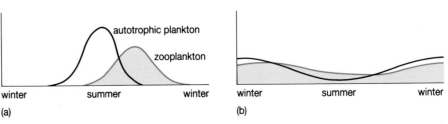

Fig 2.15 Variation in the standing crop of autotrophic plankton and zooplankton in the euphotic zone of two other latitudes.

2.6 ECOLOGICAL EFFICIENCY OF MARINE FOOD CHAINS

Let us now return to the question which we asked right at the beginning of this chapter. What determines where we catch fish? Other things being equal, which they rarely are, we would expect that the greater the primary production then the greater the secondary, tertiary and higher levels of productivity. In other words, we would expect to catch most fish where the primary productivity is greatest. However, things are not quite that simple. The energy trapped by the primary producers has to be passed down a food chain before it can be turned into fish flesh that we can eat. We need to take a quick look at this transfer process if we are going to have a complete understanding of the processes of marine production.

Secondary production

There are relatively few good estimates for pelagic secondary production. In general the pattern of secondary productivity reflects that of primary productivity, as shown by a comparison of Tables 2.1 and 2.3. Similarly, there is a lack of data for benthic secondary productivity. The greatest standing crops of the benthos occur in shallow waters, either where marine primary productivity is high or, as in estuaries, where there is an additional source of food provided by terrestrial detritus. Such areas of high benthic productivity are vital to the life of demersal fish like plaice.

Table 2.3 Secondary productivity in the sea.

Habitat	Secondary productivity (g dry weight $m^{-2} yr^{-1}$)
open ocean	8
upwelling areas	27
continental shelf	16
algal beds/reefs	60
estuaries	34

Growth efficiency

The copepod, *Calanus*, is an important herbivorous zooplankter. It filter feeds on small cells although it can pick larger cells selectively from the water. When food is abundant many of the cells consumed are only partly digested during their passage through the copepod's gut. Some of the food which is digested by *Calanus* will be used for respiration while the remainder will be converted into new *Calanus* tissues or eggs. We can define how efficiently *Calanus* produces new tissues and eggs as

$$\text{Growth efficiency} = \frac{\text{amount of herbivore tissues and eggs produced}}{\text{amount of food eaten}}$$

Since it is only the eggs and tissues of the herbivores which are available as food to carnivores the herbivore growth efficiency will be of great importance in determining how many carnivores, such as fish, we will find in a particular area. Experiments with *Calanus* show that under optimal conditions the growth efficiency can reach 30%. Under natural conditions, growth efficiencies of between 5% and 20%, with an average of 15%, are more common. Growth efficiencies for other zooplankton are shown in Table 2.4.

Ecological efficiency of food chains

At each step in a food chain some of the energy ingested by heterotrophs is lost as waste products, e.g. faeces and urine, or as heat produced during

Table 2.4 Gross growth efficiencies of marine organisms, based on field and laboratory results.

Organism	Efficiency (%)
bacteria	30–80
Oikopleura (tunicate)	4–50
arrow worms	9–50
krill	30

respiration. The longer the food chain the greater these losses will be. We can define the ecological efficiency of each trophic level in the food chain as:

$$\text{Ecological efficiency} = \frac{\text{net production of the trophic level}}{\text{net production of the preceding trophic level}}$$

For example, the ecological efficiency of the primary consumer level in Fig 2.16 is:

$$\frac{\text{net herbivore production}}{\text{net primary production}} = \frac{15}{100} = 15\%$$

So the total amount of production available as fish will depend on three factors.

1. The rate of NPP.
2. The length of the food chain down which that NPP has to pass before it is incorporated into fish large enough to catch and eat.
3. The ecological efficiency of the food chain. This varies considerably. For example, the efficiency of the transfer between autotrophs and herbivores varies between 24% in areas of food scarcity to around 3% in areas of food abundance.

The important point to realise is that only a small amount of the NPP in a particular area ends up as the fish that you and I can eat. Nonetheless, fish is an important human food resource, particularly as a source of protein, which needs careful management. The process of fisheries management forms the basis of the work in the next chapter.

QUESTIONS

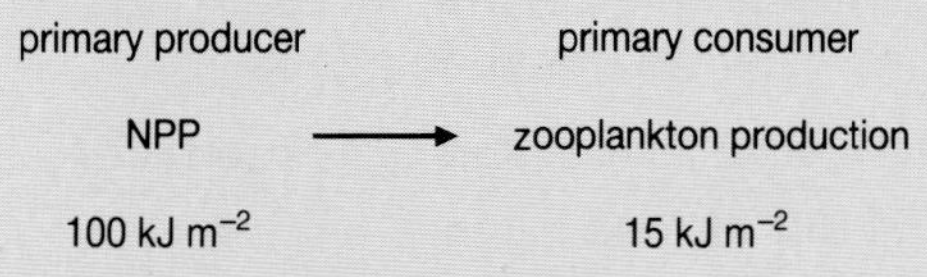

primary producer   primary consumer

NPP $\longrightarrow$ zooplankton production

100 kJ m^{-2} 15 kJ m^{-2}

Fig 2.16 The ecological efficiency of a marine food chain.

2.12 Why is the benthos of deep water so scarce even in areas of high primary productivity?

2.13 (a) Table 2.5 summarises some basic information about food chains, NPP and fishery yields in three different marine environments. Account for the difference in fish yields in the three areas.

(b) What happened to the rest of the primary production in Fig 2.16?

Table 2.5 The productivity of the sea.

Habitat	Percentage of ocean area	Primary (gC m^{-2} yr^{-1})	Trophic levels	Mean efficiency (%)	Fish production (kg $\times$ 10^9 fresh weight)
open ocean	90	50	5	10	1.6
coastal areas	9.9	100	3	15	120
upwellings	0.1	300	1.5	20	120

MARINE PRODUCTIVITY

SUMMARY ASSIGNMENT

1. Make sure that you have definitions of, and understand, the terms given in learning objective 1.

2. Why is the standing crop of autotrophic plankton in a body of water not always representative of primary productivity?

3. (a) Explain how the quantity and nature of light varies with depth in the sea. What factors affect the penetration of light?
   (b) Discuss the significance to marine plants and animals of the variations of light with depth.
   (c) Describe a simple method for measuring the clarity of sea water.

4. (a) By means of a diagram only, show the variations through the year in the biomass of autotrophic plankton and also that of zooplankton in a north temperate area.
   (b) Account for these variations, highlighting the importance of nutrient supply, turbulence, light intensity, seasons and zooplankton grazing.

5. (a) Define the terms growth and ecological efficiency.
   (b) Which will have a higher growth efficiency, a cow or a fish? Explain your answer.

Chapter 3

FISH AND FISHERIES

Fish are a valuable and important source of food with many health-promoting properties, so they need careful management to ensure that we get the maximum yield without over-exploitation. However, to achieve this management some basic understanding of both fish ecology and the fishing industry is required. In this chapter, we will concentrate on these aspects of fisheries science. In the next chapter, we will look at how fish populations (stocks) can be managed and exploited. In this book, we will confine our discussions to marine fish with bony skeletons, the teleosts, since these form the main part of the fish catch eaten in Britain. However, fresh water teleosts and those fish with skeletons made of cartilage, the elasmobranchs (sharks and rays), are widely eaten elsewhere in the world.

LEARNING OBJECTIVES

After completing the work in this chapter you will be able to:

1. describe the niches and adaptations of five major types of marine teleost;

2. outline the basic life history of temperate marine teleosts;

3. discuss the concept of **unit stock** and explain how the integrity of such stocks is maintained;

4. explain how fish can be aged;

5. account for the relationship between larval mortality, recruitment and the age class structure of fish stocks;

6. describe fishing methods and the global pattern of fish yields.

3.1 HOW FISH LIVE

We need to know where fish live and how they live, i.e. their ecological niche, since this will determine how the fish can be caught and how many fish we can catch. We can recognise five basic niches which marine teleosts can occupy.

1. **Pelagic plankton feeders** include herrings, pilchards and anchovies, which always feed near the surface, and fish like mackerel, which feed on plankton for part of the year but feed in mid-water or on the bottom at other times of the year (Fig 3.1). These fish collect plankton largely by sieving it from the water as it passes through their gill rakers, although they can select individual food items. Herrings tend to be found in more northern waters of the Atlantic and the Pacific. Pilchards are found in warmer waters, for example in the Mediterranean and off the cost of southwest Africa. Anchovies dominate this niche in the southern Pacific, for example off the coasts of Peru, Chile and southeast Asia.

2. **Pelagic predators** feed on pelagic planktivorous fish. These include tunas (see Fig 3.1), albacore, bonito and barracudas. These large fish –

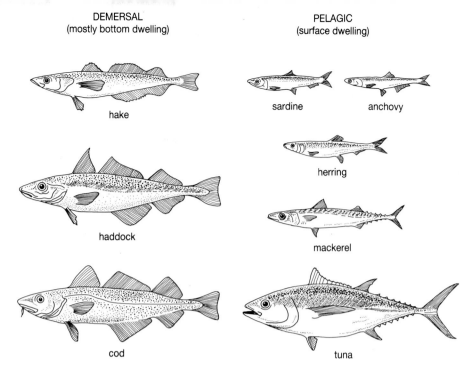

DEMERSAL (mostly bottom dwelling)	PELAGIC (surface dwelling)

hake

sardine   anchovy

herring

haddock

mackerel

cod

tuna

Fig 3.1 Some of the major types of commercially harvested marine fish. (Not drawn to scale)

some tunas reach up to 4 m in length – are commonest in warmer waters, moving into temperate waters only in the summer.

3. **Vertical migrators** feed at the surface during the night but are found at depths of 300–400 m during the day. This group includes silver fishes, like lantern and hatchet fish, which feed on plankton, and larger black fish, like *Bathophilus*, which feed on the silver fish. The black fish follow the silver fish to the surface at night.

4. **Deep-sea fishes**, found below 1000 m, are an extraordinary looking group, including deep-sea anglers, gulpers and swallowers. Although few in number this group is widely distributed.

5. **Demersal fishes** are found in the shallow seas above the continental shelves. They can be divided into two groups.
   - Those that feed both in mid-water and on the bottom, like cod, haddock and hake (see Fig 3.1), taking both small fish and benthic invertebrates.
   - Those that live entirely on the bottom, like plaice, halibut and sole (Fig 3.2), feeding almost exclusively on benthic invertebrates.

Fig 3.2 A flounder, *Platichthys flesus.* What adaptations does this fish have for living on the bottom of the sea?

QUESTION	3.1 (a) For each group of fishes described above, suggest how they might be adapted to their way of life. (b) Can you suggest how and why biomass might vary between populations of these five sets of fishes? (c) On the basis of your answer to (b) which group of fishes are likely to provide the best catches for humans?

3.2 LIFE HISTORY

Marine teleosts lay large numbers of small eggs containing little yolk (Table 3.1). The eggs are usually laid in mid-water but then float upwards to join the meroplankton. After hatching the larval fish feed on the egg yolk, which is contained in their stomachs, for a few days before starting to take planktonic food. Initially, this food is likely to be diatoms, but the

Table 3.1 Data on life histories of herring, plaice and cod in the North Sea (times and sizes are approximate because rates of growth are dependent on factors such as temperature and food supply).

Characteristic	Herring	Plaice	Cod
number of eggs laid annually by large female	2×10^4	5×10^5	4×10^6
size of egg	1.3 mm diam.	1.6 mm diam.	1.5 mm diam.
location of egg	sticking to bottom	floating	floating
days to hatching	about 10	about 15	about 17
days to absorption of yolk sac	about 14	about 8	about 7
larval habit	all three live as meroplankton		
principal larval food	copepods esp. *Pseudocalanus*	*Oikopleura*	copepods
duration of larval stage	14 weeks	5 weeks	10 weeks
length at metamorphosis	about 40 mm	about 15 mm	about 40 mm
juvenile habit	all three become demersal in shallow seas and gradually move offshore as they grow		
age at first spawning	3–5 yr	4–5 yr	5 yr
size at first spawning	about 25 mm	20–35 cm	about 70 cm
typical age of 'large' fish	>11 yr	15–40 yr	>20 yr

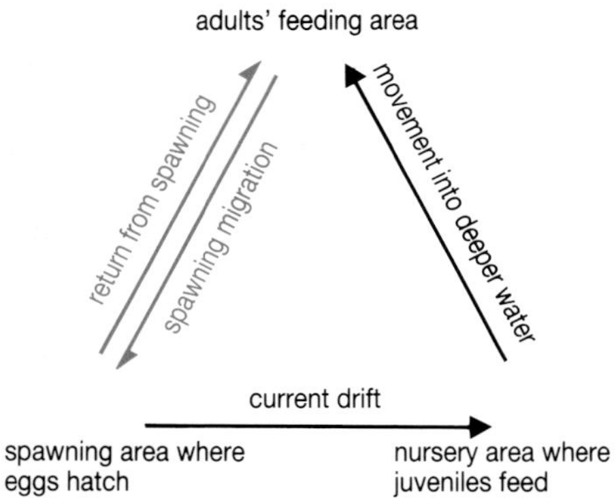

Fig 3.3 The triangular circuit of migration common to many marine fishes. Note that the feeding area for adult fish is separate from both the nursery area and the spawning area.

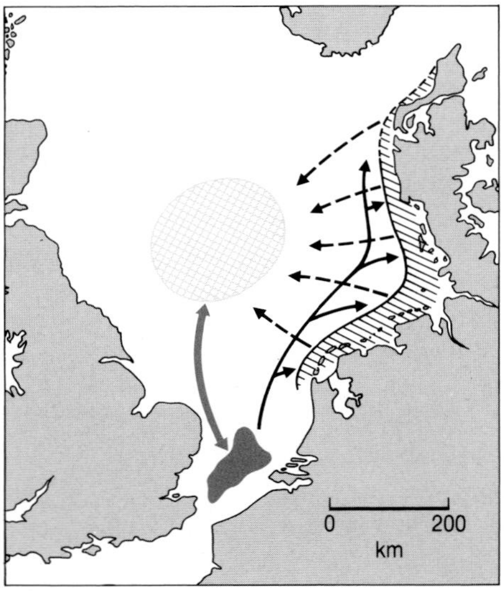

Fig 3.4 The pattern of migration for plaice in the southern North Sea. Spawning grounds are shown in red and the nursery area, the Waddensee, in cross hatching. The adult feeding grounds are shown in pink. Red arrows show the active movement of adults from feeding to spawning grounds, black arrows show the drift of larval stages to the nursery grounds, and dashed arrows show the gradual movement of juveniles into deeper water and, eventually, to the adults' feeding ground.

young fish rapidly progress on to large zooplankton, different species tending to select different zooplankters. Eventually, the larval fish meta-morphose (change) into juveniles.

The juvenile fish now start to feed in nursery areas or grounds. For continental shelf species these are found in shallow water, e.g. the Waddensee (Fig 3.4), while the juveniles of oceanic species tend to feed in shallower water than the adults. As the juveniles grow they move into deeper water until they grow large enough to join the adult fish population. They are then said to have **recruited** to the adult stock.

Since fish eggs and larvae are part of the plankton they will drift with the current, so the spawning grounds must be down current of the nursery grounds. This means that the juvenile fish will need to make a return migration back to the spawning grounds when they are ready to breed. This produces a triangular circuit of migration, as shown in Fig 3.3.

The life history of plaice, *Pleuronectes platessa*

There is a great deal of data relating to plaice from the North Sea, and their life history, which is well known, is typical of many marine teleosts. The spawning grounds of the southern North Sea plaice population lie off the Thames estuary (Fig 3.4), where the adult fish arrive in mid-January to lay and fertilise eggs. The small, yolkless eggs hatch after a few days, depending on the temperature of the sea water, into tiny planktonic larvae. The eggs and developing larvae are carried by the prevailing currents along the Belgium and Dutch coasts towards the Waddensee. Here, if the larvae have reached a certain size, they metamorphose into bottom-feeding juveniles. These young fish move into deeper water as they grow and, eventually, when they are four to five years old, they reach the adult feeding grounds (shown in pink in Fig 3.4). At this stage they have grown large enough to be caught in fisherman's nets and thus have been recruited to the adult population or **stock**.

Fish migration

Clearly, fish are capable of moving long distances, in the case of plaice up to 300 km. However, this figure is dwarfed by the migrations for species like blue fin tuna. Two individuals of this species, tagged off the coast of

Florida, were recovered 28 days later 7000 km away off the coast of Norway. Not only had they swum right across the Atlantic, but they had done so at the rate of 60 km per day. Similar long-range migrations are shown by European eels which grow to maturity in European rivers but spawn in the Sargasso Sea, 3500 km away across the Atlantic. The young eels return to Europe as planktonic passengers in the North Atlantic Drift, feeding on the rich plankton contained in this current as they go. The importance of these migratory movements, both to the biology of the fish and the fishing industry, can be seen if we examine the concept of unit stock.

The concept of unit stock

Fish species are divided into separate stocks which may have quite different characteristics. To explain how the integrity of such stocks is maintained in the apparently uniform vastness of the sea let us consider the Arcto-Norwegian cod fishery (Fig 3.5).

The cod in this stock spawn on the east side of the Lofoten Islands in middle waters during February and March. The eggs are laid at a depth of around 100 m where there is a boundary between fresher, lighter water and deeper, denser, saltier water. The adults leave the spawning grounds in April and proceed to the feeding grounds in the Barents Sea, around Bear Island, and to the southeast of Spitsbergen. Here the adults feed through the summer on benthic invertebrates and small pelagic or mid-water fish, like capelin.

Meanwhile the eggs and larvae are carried on to nursery ground in the

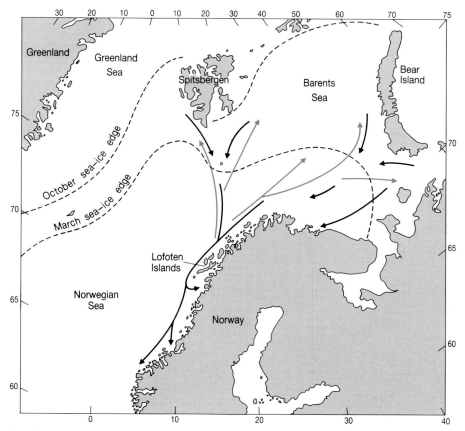

Fig 3.5 A map to show the annual migrations of the Arcto-Norwegian cod stock. The black arrows show the southward movement of the cod to the spawning grounds around the Lofoten Islands in the late winter. Red arrows show the northward movement of the spent cod from the spawning grounds in spring back to the summer feeding grounds.

Buchan
July (end)–Sept

Minch
Sept–Nov

Manx
Sept–Nov

Dogger
Sept–Nov

Downs
Nov–Jan

Fig 3.6 Spawning times of herring stocks in the Irish and North sea.

Barents Sea from where the juvenile cod recruit to the adult population. In October the cod begin to move south and east, congregating south of Bear Island and off the North Cape. Subsequently, they move south along the coast of Norway, arriving at the Lofoten spawning grounds in January. It is therefore the annual migratory cycle to spawning, nursery and adult feeding grounds which maintains the unity of the stock, isolating it from other stocks of the same species.

Different stocks within a species are usually recognised on the basis of their spawning grounds and spawning periods. An example of this is shown in Fig 3.6 for herring.

To conclude you should note the following points.

- Fish species are split into stocks. Each major stock is largely isolated from other stocks and has its own characteristics.

- The integrity of the stocks is maintained by an annual cycle of migratory movements.

- Fisheries exploit a particular stock, often at a particular stage in the migratory cycle. Thus we have the Arcto-Norwegian cod fishery, the Icelandic cod fishery and so on. Some of these fisheries can be further subdivided into spawning, summer and winter fisheries. For example, there is an established fishery for the spawning cod, called skrei by the Norwegians, off the Lofoten Islands.

- To maximise the efficiency of the fishery it is important that details of the migratory movements are known. Such information can be obtained by tagging individual fish (Fig 3.7) and recovering the tags when the fish are caught.

Fig 3.7 A tagged halibut waiting to be returned to the sea. The tag will provide information on the fish's movements when it is caught again.

FISH AND FISHERIES

• In devising management schemes the fisheries biologist needs to specify the stock which is under study, since stocks may differ in both growth rate and recruitment, for both these factors affect yield and the mass of fish taken from the stock.

QUESTIONS

3.2 **(a)** Suggest the possible genetic consequences of the division of fish species into different stocks.

(b) What evidence is there to support your suggestions?

3.3 Using Fig 3.3 as a basis, draw a simple diagram to summarise the migratory movements of the southern North Sea plaice. Indicate the process of recruitment on your diagram.

3.3 FISH POPULATION DYNAMICS

Marine teleosts produce vast numbers of eggs (see Table 3.1) but we are clearly not overrun by fish. There must therefore be enormous mortality among the fish which hatch from these eggs. Most of this seems to occur during the larval stages.

Larval mortality

This is influenced by three main factors.

1. Availability of food.
2. Temperature.
3. Predation.

If the planktonic food of the larvae is in short supply then the larvae will starve. Note, however, that because different species tend to select different zooplankters, a reduction in the density of just one type of zooplankton may only influence a specific species. Starvation may kill the larvae or, alternatively, it may slow the rate of larval development so that the fish do not grow sufficiently to metamorphose when they reach the nursery grounds. In this case the current will sweep these larvae beyond the nursery ground so that, when they do metamorphose, they are in an area where they cannot survive.

The rate of larval development is temperature dependent. Thus if the temperature of the sea is too high the larvae may metamorphose before they reach the nursery grounds; if it is too low they may metamorphose after they have been swept away from the nursery grounds. The effect is the same in both cases; juvenile fish find themselves in areas where they cannot survive.

Fish larvae form the food of a large number of predators, including other fish of the same species. High levels of predation will reduce the number of larvae reaching the nursery grounds.

Pollutants may also affect the rate at which the fish larvae develop while some species, notably sole, suffer large juvenile mortalities on the nursery grounds. All of these factors will affect the number of fish which recruit from the juvenile to the adult stocks each year.

The age structure of fish stocks

We only become aware of the variability in recruitment when members of a **year class** (fishes produced from eggs laid in the same spawning season) grow large enough to enter the **catchable stock** (the stock of fishes which can actually be caught by the fishing gear used in the fishery). When this happens we can estimate their numbers, and follow their survival, providing we can age the fish which are caught.

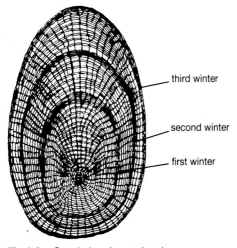

third winter

second winter

first winter

Fig 3.9 Growth rings in a cod scale.

Ageing fish

Growth rings: Most fish in temperate waters grow very little in the winter but very fast in the spring and summer. These alternating periods of slow and rapid growth are recorded in bony structures like scales, otoliths and opercular bones (Figs 3.8 and 3.9) as growth rings, just like the alternate periods of rapid and slow growth are recorded as growth rings in the trunks of trees. Counting these growth rings allows you to age the fish. However, such growth rings become less distinct with age, and some fish live a long time. Also, rings are absent in tropical species which tend to grow at the same rate all the year.

Size–frequency analysis (Petersen's method): Here the fisheries biologist records the length of a representative sample of fish. Usually, this is done using commercially caught fish on the quayside. In addition, the biologist will take otoliths and scales from the measured fish. The data on length are then used to plot length/frequency graphs (Fig 3.10). The biologist then attempts to break this complicated polymodal (many peaked) curve into its separate modes (single peaks), where each mode represents an age class. The age assigned to each age class can then be checked using otolith and/or scale markings.

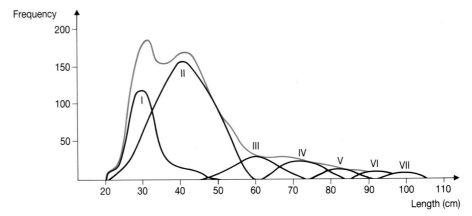

Fig 3.10 Dissection of length/frequency data for cod using Petersen's principle. Catch data are shown in red. The Roman numerals represent different age classes of fish from 1–7.

Age class strength

Fig 3.11 shows the age composition of Norwegian herring stocks caught between 1907 and 1923. From this data we can see that the year class of 1904 entered the catchable stock as four-year olds in 1908 and then dominated the catch for the next ten years. Even in 1921 the remains of this year class, by now 17 years old, are still easily observable in the overall catch.

You might think that a large year class will lay a lot of eggs, since there are a lot of fish, and you might therefore expect another strong year class to follow on. Unfortunately for fisheries managers this does not seem to happen. If you look at Fig 3.11 you will notice that after the year class of 1904 there were a number of weak year classes. This implies that recruitment must have been poor in these years despite the fact that a large number of eggs may have been laid. It appears, then, that the strength of an age class does not depend on the size of the adult stock but rather on the survival of the larvae.

It is now widely accepted that the major factors producing such variation in the strength of age classes are environmental variables, like temperature, which affect larval survival. However, such variables are not thought to affect the larvae directly but rather affect the quality and availability of larval food.

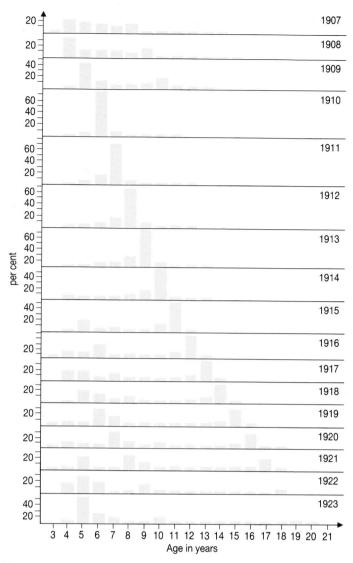

Fig 3.11 Histograms showing the percentage year class composition of Norwegian herring catches between 1907 and 1923.

The 'match–mismatch' hypothesis

Imagine a plaice egg drifting in the plankton. It hatches into a larva which requires a lot of planktonic food, particularly diatoms and zooplankton, if it is to survive and grow. Such high quality food will only be available during a plankton bloom (see Section 2.5). So there are three key events in the life of a plaice, the timing of which will affect the survival of the plaice larvae.

1. The timing of egg laying (spawning).
2. The timing of larval hatching.
3. The timing of the plankton bloom.

The relationship between these three events is summarised in Fig 3.12. The probability of larval survival will depend on the extent to which the timing of larval production coincides ('matches') with the production of larval food. This is shown in red in Fig 3.12. If there is a good **match** then larval survival will be good, which will result in a strong age class recruiting to the adult population three to four years later. By contrast, if the overlap is poor, i.e. there is a **mismatch** between larval production and larval food

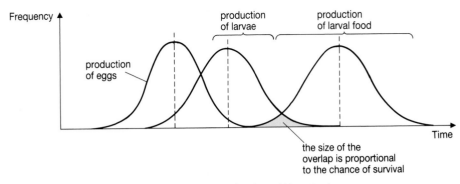

Fig 3.12 Diagrammatic representation of the 'match–mismatch' hypothesis.

supply, then larval mortality will be high, resulting in a weak age class recruiting to the adult population three to four years later.

The time of spawning can be estimated by counting the number of eggs floating in the sea; in the case of the southern North Sea plaice stock, peak spawning occurs around 19 January, a date which shows remarkably little variation from year to year. By contrast, the timing of the spring bloom in the southern North Sea can vary by up to six weeks. Consequently, there is plenty of opportunity for the production of larvae to become mismatched with the production of larval food. For example, if the spring bloom occurs later than usual many larvae will starve and their growth will be retarded. Many larvae will die as a result while those that do manage to metamorphose may well do so after they have been swept beyond the juvenile feeding grounds in the Waddensee (see Fig 3.4). Inevitably, this will result in poor recruitment to the adult population three to four years later, leading to a weak age class.

QUESTIONS

3.4 Using the information given in Fig 3.6 suggest the main sources of larval food exploited by different stocks of herring around the British Isles.

3.5 Following the cold winter of 1962–3 there was a record recruitment of adults to the southern North Sea plaice population.
   (a) When would this strong age class first appear in the fishery?
   (b) What will be the main factor determining the peak of larval production?
   (c) Suggest a hypothesis to account for the strong recruitment of plaice following the cold winter.
   (d) Why do fish not vary their time of spawning to take into account the variability in the timing of the plankton bloom?
   (e) Given the mean date of spawning for plaice is 19 January when, on average, will the larvae begin to feed on the plankton? Does this date surprise you?

3.6 (a) Using Fig 3.12 predict the effect of an increase in stock size on the degree of matching of larval production to food production. (**Hint**: increase the height and widen the bases of the first two curves.)
   (b) How therefore should such an increase in stock size affect recruitment?
   (c) Explain why the effect you predict is, in fact, not often observed. For example, the strong age class of herrings in 1908 did not produce strong age classes in subsequent years.

Fishery scientists are usually concerned with the problems generated by the fishing industry so it seems sensible that you should know something about the nature of this industry. In particular, we will look at fish yields and fishing methods.

Fish yields

For reporting purposes the United Nations Food and Agricultural Organisation (FAO) divides the world's seas into areas. Each year, data is collected from individual fisheries on, for example, the tonnage and species of fish caught, and this is published by the FAO. Such data provides a basis for comparing yields from different fisheries, from different species and for investigating long-term trends in catches. Such data and research are essential if we are going to increase the global yield of fish to the 110 million tonnes the FAO estimates will be required by the year 2000. In particular, such research enables us to predict areas where yields could be improved by either opening new fisheries or exploiting under-used species.

Fisheries data from the FAO is summarised in Tables 3.2 and 3.3. You should note the following points in these tables.

- Almost 90% of the yield is from marine organisms of which fish and molluscs (mussels, oysters and so on) are the major component.

Table 3.2 Principal fish species caught in 1984.

Species	Catch (millions tonnes yr^{-1})
Alaska pollack	5.99
Japanese pilchard	5.16
Chilean pilchard	5.02
Capelin	2.58
Chilean jack mackerel	2.41
Chub mackerel	2.21
Atlantic cod	1.97
Atlantic herring	1.19
Skipjack tuna	1.05
European sardine	0.88

Table 3.3 Catches of fish from major fishing areas.

Area	Potential catch (millions tonnes)	Actual catch
Atlantic		
N. Atlantic		13.94
Central Atlantic		7.16
S. Atlantic		3.96
Total	53.7	25.06
Pacific		
N. Pacific		26.42
Central Pacific		8.53
S. Pacific		8.68
Total	55.5	43.63
Indian Ocean	7.3	4.36
World total	118.2	82.77

- The fish catch is dominated by pelagic planktivorous species. Indeed, up to 1972 when the fishery collapsed, the Peruvian anchoveta (a type of anchovy) alone comprised 18% of the 'total global catch.
- Only a few species of fish are heavily exploited. Thus 70 out of more than 20 000 species of marine teleost constitute almost half the catch. This suggests that one way of increasing yield would be to exploit non-traditional species. Although there may be consumer resistance to, say, lantern fish and chips, such species could be used to produce fish meal or in processed foods like fish fingers. In addition to consumer resistance there is the added problem of our lack of knowledge about the biology of these lesser-used fish.

Fishing methods

The large number of different fishing techniques can be classified into three major categories.

1. Hooking individual fish, e.g. longlining and angling.
2. Tangling fish in netting, e.g. gill nets and drift nets.
3. Actively catching fish in a net, e.g. seine netting and trawling.

We will briefly look at each of these techniques.

Hook and line fishing

This is a traditional method which can be used to catch many species including cod, haddock and tuna. Hook and line fishing has largely been superseded by seining and trawling because it is so labour intensive. However, since the ships used in hook and line fisheries are small the method is still popular with inshore fishermen. Usually, some form of bait or lure is employed although mackerel can be caught on unbaited hooks since they will strike at practically anything.

The hooks are usually arranged in groups attached to a line (Fig 3.13). The lines can then be anchored to the bottom to catch demersal species such as cod or suspended from buoys to catch pelagic species like tuna. Such lines can be up to 8 km long.

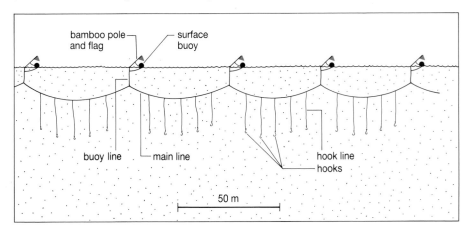

Fig 3.13 Hook and line fishing. These long lines are used to catch, for example, tuna fish over deep oceans although they are increasingly being replaced by drift nets.

Drift netting

Drift nets are essentially long curtains of netting which hang in the water (Fig 3.14). The head line is suspended from buoys while the bottom of the net is weighted by a heavy rope, the messenger, which also attaches the net to the boat. Fish are caught by their gill covers as they swim into the net as it drifts with the tides and current. Extensively used to catch herring in the North Sea and Atlantic before the Second World War, drift nets in these

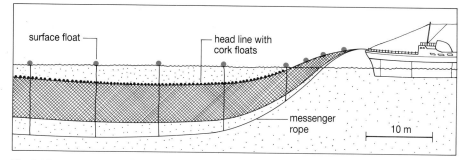

Fig 3.14 A typical drift net used over continental shelves to catch pelagic species like herring.

fisheries have now been largely superseded by more efficient purse seines and mid-water trawls. However, drift nets are still widely used in the Pacific to catch tuna where up to 30 km of nylon nets are deployed. This has led to considerable controversy since the method is so unselective and can kill marine mammals like dolphins and whales leading to a ban on these nets in late 1991.

Seining

Purse seines are widely used for pelagic shoaling species such as capelin and tuna. After a shoal is spotted, using sonar or spotting helicopters, a curtain of net, up to 1.5 km long and 100 m deep, is 'shot' (launched) around the fish (Fig 3.15). Once the shoal is surrounded a line running around the bottom of the net, the purse line or seine ring, is hauled in, so closing the bottom of the net. The fish, now trapped in a purse of netting (hence purse seining), are concentrated as the net is slowly hauled in and the net purse becomes shallower. Eventually, the fish become so concentrated that they can be scooped or, in the case of small species, pumped on board the ship.

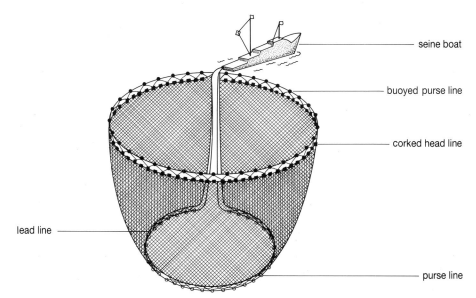

Fig 3.15 A purse seine. This type of net is widely used to catch pelagic species like anchovy, herring and mackerel.

Danish seines are lightweight nets used for catching fish on smooth sea bottoms. Since they are small and manageable they can be deployed from small boats in shallow water. The net consists of a central sock with wings of net on either side of the central opening. The ground rope is weighted while the head rope is buoyed, so keeping the net open. This net can be

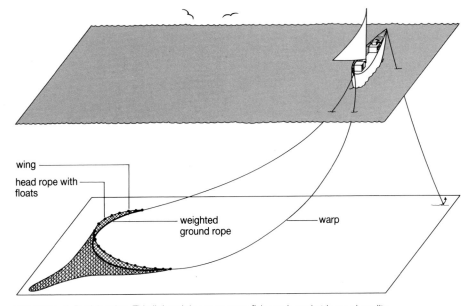

wing

head rope with floats

weighted ground rope

warp

Fig 3.16 A Danish seine. This lightweight net ensures fish reach market in good quality.

used in a number of ways, one of which is shown in Fig 3.16. The fish landed using this technique are usually in good condition and can often be delivered to the fish market while they are still alive, so commanding a better price.

Trawling

A trawl is a sock of net which is pulled through the water. When the fish first meet the advancing trawl they swim ahead of it. However, provided the trawl is towed fast enough the fish soon tire and fall exhausted back into the mouth of the net.

Two types of trawl are most widely used. In the beam trawl the mouth of the net is kept open by a stout beam about 16.5 m long (Fig 3.17). This trawl is still widely used in the Dutch sole fishery and in prawn fisheries. The efficiency of the trawl can be improved by the addition of tickler chains to the beam. These disturb the sea bed, so frightening fish which, otherwise, could lie still in the sediment and let the net pass over them.

The beam trawl is unwieldy to use and was superseded by the beginning of this century by the otter trawl. Here the mouth of the trawl is kept open by the use of otter boards (Fig 3.18). These are large wooden or steel rectangles about 3×1.5 m in size, weighing around 1 tonne. An otter board

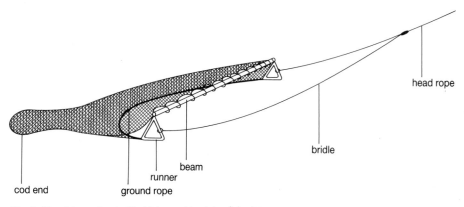

head rope

bridle

beam

runner

ground rope

cod end

Fig 3.17 A beam trawl still widely used in plaice fisheries.

FISH AND FISHERIES

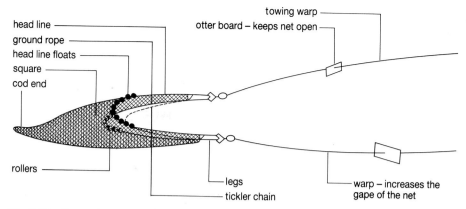

head line
ground rope
head line floats
square
cod end

towing warp
otter board – keeps net open

rollers

legs
tickler chain

warp – increases the
gape of the net

Fig 3.18 The modern otter trawl requires large, powerful boats to pull it across the sea bottom.

is attached to each trawl arm. The important point to note is that the inside of each otter board is attached to the towing warp (rope) just forward of its centre. This arrangement forces the otter boards outwards as the net is pulled through the water, so opening the mouth of the net. The addition of a piece of wire between the otter boards and net allows the mouth of the net to be spread even further while the wires disturb the fish over a larger area, so increasing the efficiency of the net. Such **Vigneron Dahl** gear was introduced in the 1920s.

A major advance in trawler technology was the introduction of stern trawlers. Unlike side trawlers, which must turn sideways on to the waves as they recover their nets, a stern trawler can shoot (let out), and recover, its gear by using a ramp at the stern of the ship. Such trawlers can therefore head into the waves when retrieving their gear and so can operate under stormier conditions than side trawlers.

As the trawl is retrieved the catch becomes concentrated in the cod end of the net which is closed by a draw string or cod line. Fish are removed through the bottom of the cod end by undoing the cod line. The fish, usually in poorer condition than those caught using a Danish seine, are then sorted and packed in ice to keep them fresh.

Bottom trawls have now been adapted for use in mid-water fisheries such as the krill fisheries in the Antarctic. These nets are immense, up to 100 m long with a mouth opening of 400–500 m². The otter boards attached to these nets can weigh 3 tonnes each and they require large, powerful trawlers for their operation.

QUESTIONS

3.7 Table 3.3 shows the catches from major fishing areas for 1986.
   (a) Which ocean produces the most fish?
   (b) Which ocean produces the most fish per unit area?
   (c) Why is your answer to (a) not the same as your answer to (b)?
   (d) Which ocean has the greatest potential for increasing its catch?

3.8 (a) Explain why the advent of steam-driven ships increased the efficiency of trawling as a means of catching fish.
   (b) Suggest the purpose of (i) the rollers (ii) the head line floats on the trawl shown in Fig 3.18.

3.9 (a) Why does the vast majority of the fish caught by the world's fishermen consist of pelagic fish which form huge shoals, e.g. herrings, sardines and pilchards?
   (b) Why might catching a widely dispersed, non-shoaling fish not be an attractive proposition for a very large commercial fishery?

FISH AND FISHERIES

1. Keep a copy of your answer to Question 3.1 to remind you about the main types of marine teleosts.

2. Explain how the annual cycle of migration shown by temperate marine fishes maintains separate unit stocks.

3. **(a)** Outline three methods for ageing fish.
   (b) Why do age class strengths vary from year to year?
   (c) Suggest why information on age class strength is vital for fisheries management.

4. Compare and contrast the following pairs of fishing methods:
   (a) drift netting and purse seining
   (b) Danish seining and trawling.

5. 'It is vitally important for mankind to gain protein from the oceans ... and to realise that potential marine productivity is finite' (Russel-Hunter 1970). Research and discuss this.

Chapter 4

HARVESTING THE SEA 1: FISHERY MODELS

Harvesting fish involves treading a narrow path between over-exploitation and under-exploitation. If a fish stock is overfished it will, inevitably, decline in size and may ultimately reach such a small size that it can no longer provide a harvest. Underfishing means that we could actually take more fish from the stock without damaging future stocks. This would provide both more protein for hungry mouths and jobs in often remote communities which rely on fishing as their main source of income. In this chapter, we will examine how fisheries managers try to stay on this narrow path.

Fig 4.1 Peterhead Harbour, a major fishing port in North East Scotland. Recent changes to fish quotas in the North Sea could cause busy ports like this one to suffer a severe economic decline.

LEARNING OBJECTIVES

After completing the work in this chapter you will be able to:

1. use a simple model and an equation to explain how the biomass of an unexploited fish stock changes;

2. describe the relationship between stock size, fishing effort and sustainable yield;

3. explain the difficulties of obtaining reliable information about fish stocks;

4. discuss the applications and limitations of two types of surplus yield fishery models: constant quota and constant effort.

4.1 FISH POPULATION DYNAMICS

The central problem of fisheries management can be stated in the form of a question: How many tonnes of fish can be harvested from this particular stock this year so that when we go back next year we still catch enough fish? This idea of being able to take a harvest, a **sustainable yield**, from the same stock of fish year after year is central to fisheries management. However, fishing is a commercial business; it has to make money. So what the fishermen need to know is what is the **maximum sustainable yield (MSY)**, the maximum mass of fish they can catch in any one year without impairing the harvest from the fishery in future years.

Modelling stock growth

To understand how a fishery can be regulated we first need to consider the factors which determine the size of a stock of fish when it is not being exploited. To gain this insight, we will use the simple model shown in Fig 4.2. The model is not supposed to be realistic. Rather it is designed to help you to think about and understand how a stock can increase and decrease in size, and how it can be fished without damaging future yields. Since fishery managers are primarily concerned with the mass, and not the numbers of fish landed, we will model changes in the biomass of the stock.

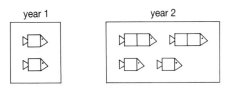
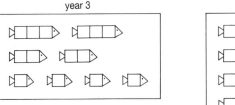
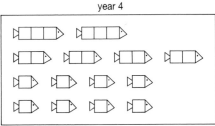

Fig 4.2   A simple model to show the increase in the biomass of a stock of blockfish.

Running the model

The model in Fig 4.2 shows the growth in a stock of fish over four years. In the first year we start with two fish, one male and one female, each of which weighs one unit of biomass. Thus in year 1 the stock size is two biomass units. By year 2 both the original fish have grown, adding another unit of biomass each. In addition, they have produced sufficient fertilised eggs for two further fish to recruit to the population, each with one unit of biomass. So in year 2 the stock size is six units of biomass. The increase of four units of biomass has resulted from two processes: growth (G) and recruitment (R). By year 3 the two original fish have grown by a further unit of biomass, as have the two recruits which joined the population in year 2. In addition, a further four recruits have joined the population so the total stock size is now 14 biomass units. Again, the stock has grown by a combination of growth and recruitment. By year 4 the two original fish have died (blockfish only live three years). Other fish present in the stock in the previous year have grown by one unit of biomass and eight new recruits have joined the population. So the total biomass is now 22 units.

The last year of our model, year 3 to 4, shows that the stock size in year 4 depends on:

- the stock size in the previous year (S);
- biomass added by growth of fish already in the stock (G);
- biomass added by recruitment (R);
- biomass lost by mortality (M).

We can combine these terms into a fundamentally important equation:

$$S_n = S_{n-1} + G + R - M$$

where S_n is the stock size now and S_{n-1} is the stock size last year.

The rate of growth of a fish stock

This exercise will give you practice in data analysis, and graphical and mathematical skills.

Continue running the model shown in Fig 4.2 for a further four years. Remember three-year-old blockfish die after the females have spawned. Assume that half the fish in the stock are female and that each female lays sufficient eggs to provide two new recruits to the population in the following year.

(a) Plot the increase in biomass of the stock from year 1 to year 8.

(b) How does the rate of growth in the stock change with time?

(c) How does the rate of growth of the stock change with stock size? Try and draw a graph to show this.

(d) In what ways are the assumptions we are making in this modelling process unrealistic?

Stock growth

The analysis of the rate of growth of a fish stock shows a stock which grows at an ever increasing rate (Fig 4.3(a)). However, stocks cannot grow to an infinitely large size. Eventually, the growth rate will slow down, producing a pattern of stock growth like the one shown in Fig 4.3(b). Here the stock reaches some equilibrium size (B_{max}), where the size of the stock remains the same from year to year, i.e. $S_n = S_{n-1}$. This is an **equilibrium** situation where additions to the stock biomass through growth and recruitment are exactly balanced by losses due to mortality, i.e. $G + R = M$.

Fig 4.3 **(a)** Exponential growth.
(b) Sigmoid or logistic growth.

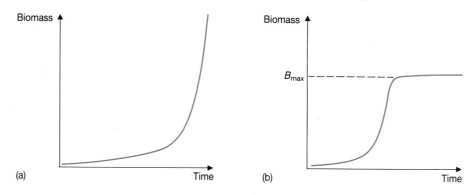

4.1 Suggest some of the factors which cause the stock growth rate to slow down, producing the curve shown in Fig 4.3(b).

4.2 Write an equation which predicts the stock size next year (S_{n+1}) if you know the stock size this year (S_n).

4.3 **(a)** What will happen to growth, recruitment and mortality in the fish stock shown in Fig 4.3(b) if stock size (i) increased above (ii) decreased below B_{max}?

(b) How would these changes in growth, recruitment and mortality affect the stock size?

4.2 HARVESTING THE STOCK

We are now ready to think about how we could harvest a stock which grows like the one shown in Fig 4.3(b). Remember that our aim is to achieve the maximum sustainable yield (MSY): the maximum tonnage of fish which can be removed from the stock year after year without damaging it.

Basic harvesting strategies

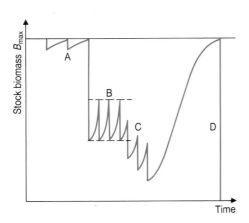

Fig 4.4 Four strategies you could use to harvest a stock of fish. Only A and B provide a sustainable yield. Strategy B provides the maximum sustainable yield because you are harvesting the fish stock at a size where it grows the fastest.

Look first at harvesting strategy A shown in Fig 4.4. Here we allow the stock to grow back to its original size before we take another harvest. The harvests are not large compared with strategy B. Here a series of regular crops keep the biomass oscillating between the levels indicated by the dotted lines. In strategy C, we start to harvest the population before it has recovered, i.e. the stock is being harvested before we have allowed enough time for the biomass to recover and so the total stock size declines. We stop fishing and the biomass recovers until it reaches the carrying capacity again. Now we take one enormous harvest, D, which exterminates the stock.

Here then we have four basic strategies. A and B both provide sustainable yields while C does not. D provides the maximum yield but it is not sustainable. The best strategy is clearly B since this provides the maximum sustainable yield. What we now need to know is the level of stock biomass and **fishing effort** which will give the maximum sustainable yield. To answer this question, we need to look at stock growth in a slightly different way.

The concept of surplus yield

Imagine a stock of blockfish like the one shown in Fig 4.5. During the course of one year the stock increases by 14 units of biomass. Now if we remove those 14 units of biomass by fishing (10 units) and natural mortality (4 units) the blockfish stock returns to where it started from. It grows and once again adds a further 14 units of biomass, from which once again we remove ten units by fishing. In other words, these ten biomass units, the difference between growth $(G + R)$ and death (M), represent surplus yield, which we can remove by fishing without damaging the stock. Since we can take the same biomass each year the ten biomass units also represent a sustainable yield. We are harvesting this stock using strategy B in Fig 4.4.

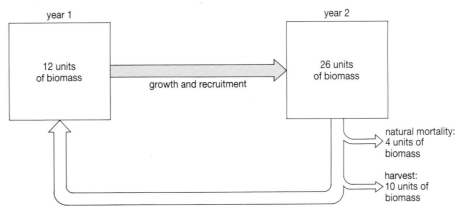

Fig 4.5 The concept of surplus yield. You can harvest ten units of biomass from this blockfish stock for ever, providing environmental conditions do not alter, so affecting growth, recruitment or natural mortality.

ANALYSIS

Surplus yield depends on stock size
This is a data-analysis exercise. You will need the results from your previous analysis of stock growth to complete it.

(a) Calculate the surplus yield for each of the stock densities shown in Fig 4.2 and the stock densities which you calculated using the model.
(b) Express your results in the form of a graph.
(c) How does surplus yield change as the stock size increases?
(d) Are all the surplus yields you calculated sustainable yields?

HARVESTING THE SEA 1: FISHERY MODELS

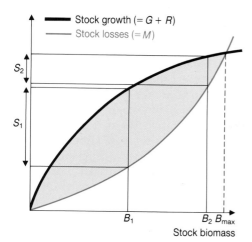

Fig 4.6 Stock growth can be seen as resulting from additions (shown in red) and losses (shown in black). Surplus yield at different stock densities is shown in pink. At biomass B_1, surplus yield is S_1, at B_2 it is S_2 and so on.

The analysis of surplus yields should show that surplus yield increases as stock size increases. However, stocks do not keep growing at the same rate for ever. Eventually, the growth rate slows down and the stock comes to an equilibrium size (see Fig 4.3(b)). We need to know how this changing pattern of stock growth affects surplus yield at different stock sizes. Fig 4.6 shows how additions to the stock (black line) and losses from it (red line) change with stock size for a stock growing like the one in Fig 4.3(b). The pink shaded areas in Fig 4.6 represent **surplus yield**.

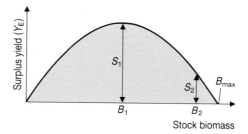

Fig 4.7 A surplus or equilibrium yield curve. This shows that the rate at which new biomass is added to a stock at different stock sizes. Use S_1 and S_2 to see the link between this graph and Fig 4.6.

Now there must be some stock size where the surplus yield is at a maximum, i.e. a maximum sustainable yield. We can see what this value is by replotting the data in Fig 4.6 to produce a graph like Fig 4.7. Here we have plotted surplus yield (Y_E) against the stock biomass. Clearly, as the stock size increases, surplus yield starts small, gets larger, rises to a peak, declines and, eventually, becomes negative (losses exceed gains at these high stock sizes). These **surplus yield curves** or **harvest parabolas** form the basis of what are called **surplus yield fishery models**. Such models are commonly used to manage fisheries and we will examine them in more detail later. However, a fisheries model is only as good as the data on which it is based, so we first need to consider the sort of information a fisheries biologist needs to construct curves like those in Fig 4.7.

QUESTIONS

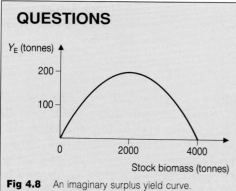

Fig 4.8 An imaginary surplus yield curve.

4.4 (a) Rewrite the equation on page 48 so that it includes mortality due to fishing.
   (b) If a yield is to be sustainable what must the relationship be between the rate of increase in stock biomass due to ($G + R$) and the rate of loss due to ($M + F$)?

4.5 (a) Fig 4.8 shows a surplus yield curve. What is the sustainable yield when the stock size is (i) 1000 tonnes (ii) 3000 tonnes?
   (b) What biomass gives the maximum sustainable yield? How is this related to the maximum biomass (B_{max})?
   (c) Why would the surplus yield curve become negative at biomasses above B_{max}?

4.3 GATHERING FISHERIES DATA

To construct a harvest parabola, we need information on:

- stock size;
- biomass increase due to stock growth (G);
- biomass increase due to recruitment (R);
- biomass losses due to mortality, as a result of both natural causes (M) and fishing (F).

Fig 4.9 A fishery inspector measuring the length of fish from a recent catch landed on a quayside. This ensures that the fish caught are above the minimum size whilst also providing essential data on the status of the fish stock.

Sampling

To estimate these parameters, we first require a sample, and the most obvious sample to use is the catch of the fishermen (Fig 4.9). The exact composition of this catch will depend on the type of gear used and the mesh size of the nets. If there are no restrictions on, for example, mesh size, then a commercial catch may include all the size classes of the stock. However, nets are usually designed to allow small fish to escape, so the commercial catch only provides information about the catchable fish. This data can then be supplemented by data obtained from special fishery research vessels using gear which allows all the size classes, including the planktonic larval stages, to be sampled. For example, fishing grounds may be sampled using trawls with a fine mesh net around the cod end, which catches even the small fish, while pelagic fish can be sampled using drift nets with different size meshes.

Such a sample would enable the fisheries biologist to age the fish and to assess the mean mass of fish in each age group, so enabling stock growth rates and hence G to be calculated. Total mortality can be assessed by noting the number of fish in each age class and how this changes over the years. Examining commercial catches enables fishing mortality to be assessed. The natural mortality can then be calculated by subtracting fishing mortality from total mortality.

Recruitment can be estimated by calculating the number of eggs that could be laid by the females in the stock. This will largely depend on their size. However, as we saw in the last chapter, it is larval survival, which depends primarily on environmental factors like temperature and food availability, rather than the number of eggs laid which is crucial in determining recruitment. Gathering detailed information on larval survival is both difficult and expensive, involving the taking of repeated plankton samples and the identification and counting of the different fish species in the sample. However, such information can be invaluable since it may allow fisheries biologists to predict when there will be good and bad fishing years for particular species, allowing the fisheries managers, potentially at least, to alter the harvesting strategies to conserve stocks in poor years and exploit them more heavily in good years. For example, the Scottish Fisheries Board noted unusually large numbers of haddock larvae in the plankton in 1922 and correctly predicted good fishing three or four years later; they were able to repeat this prediction in the late 1960s.

Tagging experiments

Fish can be captured and fitted with tags (see Fig 3.7) and then released back into the sea. Such tagged fish, when they are recaptured, provide invaluable data on growth rates, movements and fishing mortality. For example, if out of a sample of tagged fish 40% are recaptured, then fishing mortality can be estimated as 40% of the stock. This technique enabled an estimate of the fishing mortality of North Sea plaice prior to 1914 to be made at 70% of the stock. Since annual landings of this species amounted to about 50 000 tonnes, the total stock biomass must have been about 70 000 tonnes, i.e. 50 000/0.7. With about 3000 fish per tonne this gives an average stock size of some 2.1×10^8 fish.

ANALYSIS	**Using egg counts to estimate stock size**
	This is a data-handling exercise.
	Using data from plankton trawls it is possible to make an estimate of the total number of eggs laid by a particular species in a given spawning season. If you combine this with data on the average number

of eggs laid per female, you should be able to calculate the number of females in the stock and, assuming a 1:1 sex ratio, then calculate the total stock size. Such an approach was used by H.J.B. Wollaston to estimate the size of the southern North Sea plaice population in 1914, a particularly good spawning year. In that year Wollaston estimated that total egg production by this plaice stock was 3.5×10^{12} eggs.

(a) Given that the mean number of eggs laid per female is 70 000 calculate the number of females in this population.

(b) Assuming the sex ratio is 1:1 calculate the total number of individuals in this stock.

(c) How does this value compare with that calculated using the data on fishing mortality of tagged plaice?

Catch per unit fishing effort as an estimate of stock size

Imagine a population of fish. When the population is large it will require very little effort to catch the fish; when the population is small a lot more effort will be required to catch the same number of fish. We can measure fishing effort in terms of the time spent fishing, say the number of days a boat or fishing fleet spends at sea, e.g. trawler tonne days. Now, if we know the size of the catch and the amount of effort expended in obtaining that catch, we can calculate catch per unit effort. This ratio is an index of population size. Since this index is easily calculated using data from commercial catches it is often used as an estimate of stock size by fisheries biologists.

4.6 When estimating variables such as growth rate, recruitment and mortality it is important that the fisheries biologist takes a random sample from the stock.
(a) Why is a random sample needed?
(b) Will a commercial catch, after landing on the quayside, be a random sample of the stock? Explain your answer.

4.7 One way of estimating population size is to use a mark–capture–recapture method like the Petersen index:

$$\text{Population size} = \frac{\text{(number captured and marked in first sample)} \times \text{(number captured in second sample)}}{\text{number marked in second sample}}$$

Table 4.1

Sample	Number of fish caught	Number marked
1	1720	1720
2	2764	873

(a) Use the data in Table 4.1 to estimate the population size using the Petersen index.
(b) What is the fishing mortality suffered by this stock?
(c) How could you convert your population estimate into stock size in tonnes?

4.4 MANAGING FISHERIES

In this section, we are going to examine the methods which can be used to regulate fisheries and so prevent overfishing. To do this we will need a model which we can use to analyse changes in stock size in the fishery. The basis of this model will be the surplus yield curve we looked at in Section 4.2. Fisheries models based on such surplus yield curves are called **surplus yield models**. Such a surplus yield model can then be used by fisheries managers to predict the stock biomass and, more importantly, the fishing effort needed to provide the maximum sustainable yield.

Surplus yield models

The stages in the development of a surplus yield model are shown in Fig 4.10. We have assumed that the stock generates biomass according to an S-shaped (sigmoidal) biomass generation curve (Fig 4.10(a)). This gives the surplus yield curve shown in Fig 4.10(b)). Note that with our sigmoidal increase in biomass the maximum sustainable yield occurs at half the maximum biomass of the fishery. In Fig 4.10(c) we have assumed that the relationship between **catch per unit effort (U_e)** (which is an index of stock size) and the fishing effort is a straight line. The fewer fish there are the more effort it requires to catch them. We can now plot equilibrium yield against fishing effort (Fig 4.10(d)) and this gives us the **optimum fishing effort (f_{opt})** which will produce the maximum sustainable yield. Fig 4.10(d) also shows that as we increase the fishing effort beyond f_{opt} (the area shaded pink) the yield from the fishery falls, because the stock is being **overfished**. When the effort is less than f_{opt} the fishery is being **underfished** (the area shaded grey).

When interpreting surplus yield curves it is essential that you pay careful attention to the values being plotted on the x-axis. This will be either biomass or fishing effort. Effort can be given in a variety of units including percentage of stock caught, trawler tonne days and so on.

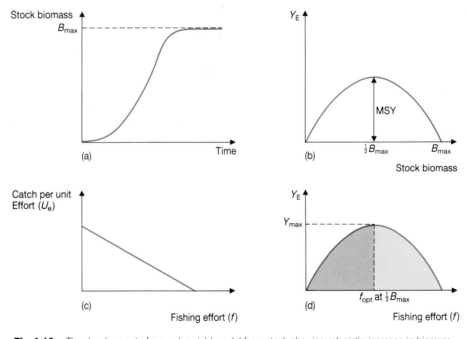

Fig 4.10 The development of a surplus yield model for a stock showing a logistic increase in biomass. See text for further details.

ANALYSIS

Interpreting surplus yield curves
This is a data-analysis exercise requiring the use of graphical interpretation skills.

Fig 4.11 shows some data from the yellowfin tuna fishery in the eastern Pacific. The red line is the equilibrium yield. The numbers are the years in which the catches were made and they are joined in sequence.
(a) What is being represented on the x-axis?
(b) Suggest how the fisheries scientists produced this curve.

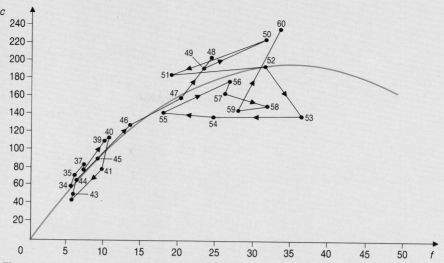

Fig 4.11 Data from the yellowfin tuna fishery in the eastern tropical Pacific. Here the catch, c, is in 10^6 lb yr^{-1} and fishing effort, f, is in 'annual standard fishing days' $\times 10^3$. The red line is the estimated surplus yield at different levels of fishing intensity. The numbers are the years in which catches were made (1934–1960) and are joined in order.

(c) Using Fig 4.11 estimate the sustainable yield of yellowfin tuna for fishing intensities of 10×10^3 and 25×10^3 'annual standard fishing days'.

(d) What would be the catch per unit effort for each of these fishing intensities?

(e) What is the optimum fishing intensity for this fishery which would give the maximum sustainable yield? What is the catch per unit effort?

Clearly, then, the fishery manager, using surplus yield models to control fishing activity in a particular fishery, is treading a thin line between overfishing, which would send the stock to extinction, and underfishing, where less fish is being taken than the stock can sustain. The manager therefore needs harvesting strategies which will allow the maximum sustainable yield to be taken from the fishery. Such strategies need to ensure that repeated and regular harvests can be taken from the stock while leaving behind sufficient individuals to breed and grow, so providing biomass for future harvests.

QUESTION

4.8 Fig 4.12 shows the relationship between fishing effort (f) and catch per unit effort (c/f) for the Pacific halibut industry.
  (a) How is the fishing effort related to stock size?
  (b) What was happening to the stock size between 1925 and 1930?
  (c) Why was this change occurring?
  (d) What happened to the fishing effort from 1930 to 1936?
  (e) How did this affect catch per unit effort?
  (f) What does this suggest is happening to the stock size?
  (g) Comment on the state of the fishery from 1936 to 1947.

4.5 FIXED QUOTA AND FIXED EFFORT HARVESTING

In this section, we are going to examine two harvesting strategies, **fixed quota harvesting** and **fixed effort harvesting**, which can be used to manage a fishery so that it provides the maximum sustainable yield.

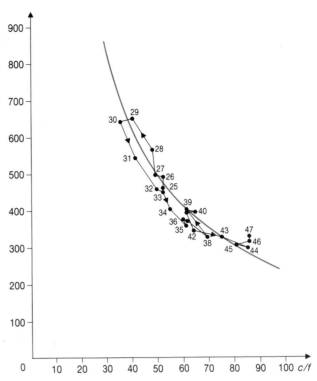

Fig 4.12 Data from the Pacific halibut fishery. f is fishing effort in annual standard fishing days. c is catch in tonnes per year. The numbers are the years in which catches were made.

Fixed quota harvesting

In fixed quota harvesting a constant biomass is removed from the stock each year. The possible consequences of this are shown in Fig. 4.13. This figure requires careful interpretation. On the x-axis, we have stock size in units of biomass. On the y-axis, we have plotted two variables, surplus yield and harvesting rate. The equilibrium yield curve is shown in red. Two possible harvesting rates, i.e. quotas, are shown in black. Remember that the harvesting rate is the biomass which is being removed from the stock each year, while the surplus yield curve shows what is being added to the stock each year, at different equilibrium biomasses, by growth and recruitment.

Consider first harvesting rate 1, the low quota. Imagine we start fishing the stock which is at its carrying capacity, B_{max}, the equilibrium stock size in the absence of fishing. Initially, the harvesting rate is greater than the equilibrium yield; the black line is higher than the red curve. So the stock decreases in size as shown by the red arrow. Eventually, we come to a point where the harvesting rate = the equilibrium yield, i.e. where the red and black lines cross. Now the harvesting rate exactly equals the stock growth rate, i.e. we are removing biomass from the stock at the same rate as the stock is replacing it. This gives us a new equilibrium stock size, B_1, from which we can take a sustainable yield. Clearly, though, this fishery could support much larger catches, i.e. we could fix a higher quota.

Look now at harvesting strategy 2, the high quota. Here we are removing from the population far more biomass than the stock can replace in a year, i.e. the harvesting rate exceeds the stock growth rate. We are overfishing the stock and it will inevitably decrease in size, shown by the little red arrows, until, eventually, it becomes extinct.

The dangers of fixed quota harvesting

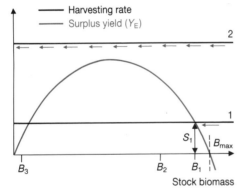

Fig 4.13 Fixed quota harvesting. (See text for details of how to use this model.)

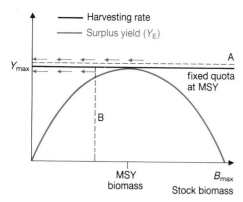

Fig 4.14 The dangers of fixed quota harvesting.

Fig 4.15 A grey whale, one of the large baleen whales, breaching (leaping out of the water) in the calving and mating grounds off the coast of Baja, California.

for the two reasons shown in Fig 4.14. Consider first the situation labelled A. Here we have slightly overestimated the MSY. This is highly likely given the nature of the variable world in which fish live. This harvesting strategy removes biomass at a faster rate than it can be replaced by the stock. The stock declines in size but we continue harvesting at what we *think* is the MSY. In the second year the difference between the harvest and the replacement of biomass is even greater, so the stock declines even further in size. This process will continue until the stock biomass declines to zero.

Now consider situation B. Here we have correctly estimated MSY but we have hit a year when the stock size, for some reason, is slightly below the equilibrium MSY biomass. This could be due, for example, to poor recruitment. The harvesting rate again exceeds biomass replacement and the stock is once again on the slippery slope to extinction.

Despite the dangers of fixed quota harvesting it has been widely used. In any particular year the MSY is estimated, the fishery opens and remains open until the quota is filled, and then closes. An obvious example of such a harvesting strategy can be found in the whaling industry. Although not fish, the fate of the large whales (Fig 4.15) starkly illustrates the dangers of fixed quota harvesting. Between 1949 and 1960 annual quotas were imposed by the International Whaling Commission. During this period whaling was a free for all and the ensuing scramble to catch the whales, before the quotas were used up, was referred to as the whaling olympics. After 1960 the whaling nations agreed to give each other quotas but even so the effects of fixed quota harvesting on whale populations are only too obvious (Fig 4.16). Another example of fixed quota harvesting is to be found in the Peruvian anchoveta industry which is discussed in more detail in the next chapter.

A slightly more sophisticated approach is neither to set quota at the MSY levels nor to leave them unchanged from year to year. For example, in the southern North Sea plaice fishery the quota is 112 000 tonnes per annum, but this is revised downwards if recruitment is poor. Such an approach does, however, require good population dynamics data which is expensive to collect and is usually not available.

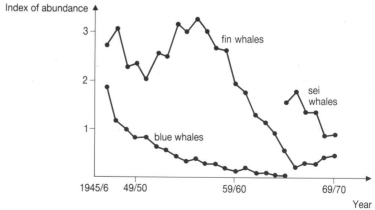

Fig 4.16 The decline in the abundance of Antarctic baleen whales under the influence of human harvesting.

Fixed effort harvesting

An alternative to fixing a quota is to fix the fishing effort. The yield from a fishery (H) depends on three things.

1. The stock size (S).
2. The level of fishing effort (e.g. the number of trawler tonne days in a fishery) (E).

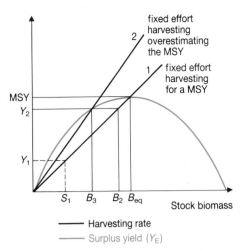

Fig 4.17 Fixed effort harvesting. (See text for details of how to use this model.)

3. The harvesting efficiency (G). This will depend on the type of gear being used, sizes of boats and so on.

Then $H = G \times E \times S$.

Assume harvesting efficiency stays the same, i.e. there are no major technical improvements in fishing gear or techniques, and we fix the fishing effort. Then the harvest is proportional to the stock size, i.e. there is a straight line relationship between them as shown in Fig 4.17. Again, this graph requires careful interpretation. We have plotted stock size on the x-axis with surplus yield and harvesting rate (yield from the harvest) on the y-axis. The equilibrium yield curve is shown in red and we have two harvesting rates shown in black. Remember the harvesting rate is simply proportional to stock size. So at stock size S_1 the yield from harvesting strategy 1 is Y_1.

Consider first harvesting strategy 1 which has set the effort to achieve the MSY; there is an equilibrium between harvesting rate and equilibrium yield which gives the MSY, providing the stock is at the equilibrium biomass, B_{eq}, which gives the MSY. Now if the stock biomass decreases below B_{eq} to B_2 then the harvesting rate falls to Y_2, which is below the rate at which biomass is replaced by the stock (the black line is below the red curve in Fig 4.17). The stock will therefore produce more biomass than is removed by fishing and will increase in size back towards B_{eq}. By contrast a fixed quota system under the same circumstances would have driven the population to extinction (case B in Fig 4.14).

The other way we could drive the stock to extinction under a fixed quota harvest system was to overestimate the MSY. This is precisely what harvesting strategy 2 in Fig 4.17 is doing. But this time, with fixed effort harvesting, rather than driving the stock to extinction we will get a new equilibrium stock size, B_3, providing a sustainable yield Y_2. In fact, we have to make a considerable overestimate of MSY effort before the stock is sent on the road to extinction.

The cost of fixed effort harvesting

There is a price to be paid for the inbuilt safety of fixed effort harvesting. The amount of fish you catch, the yield, varies with the stock size. In other words, just because you had a good year this year does not mean you will have a good year next year. Fixed effort harvesting therefore needs enforceable regulations and fishery patrol vessels to ensure that the rules are complied with and that fishermen do not increase their effort. For example, effort could be regulated by controlling the size of the fleet, or the type of vessels in the fleet. In the Pacific halibut fishery effort is controlled by seasonal closures of the fishery, and sanctuary zones where fishing is banned. This requires considerable investment in fishery protection vessels. In December 1990 a reduction in effort was introduced into the North Sea haddock and cod fishery by making all fishing vessels remain in harbour for eight consecutive days each month.

Fishery economics

All the arguments we have put forward so far about regulating fisheries have been biological ones. Our concern has been to get the best return from the stock, in terms of the maximum catch, without destroying the fishery. However, fishing is a business so perhaps we should think not so much in terms of return, the size of the catch, but in terms of its profitability. To do this we need to take into account the costs incurred by fishermen, primarily the interest on loans used to buy boats and the price of oil, as well as the selling price of the fish. Again, we can use a graphical model to explore where the optimum economic yield might be (Fig 4.18). Now the harvest-

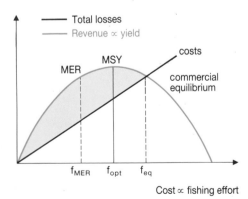

Fig 4.18 An economic fishery model. Here the maximum economic return (MER) depends as much on costs as biology. Note, we are assuming revenue from fishing is proportional to yield and the cost of fishing is proportional to fishing effort.

ing rate depends not just on the MSY but also on the costs to the fishermen. The fishing effort to achieve commercial equilibrium lies at f_{eq}. But the maximum economic return (MER) to the fishermen is where the distance between the black cost line and the red revenue curve is the greatest, i.e. when the fishing effort is at f_{MER}, i.e. below f_{opt}. In other words, it pays the fishermen, in this case, to underfish.

QUESTIONS

4.9 (a) Imagine a stock at B_2 in Fig 4.13. What would happen to such a stock if you harvested it according to strategy 1, the low quota? Explain your answer.

(b) What would happen to stock at B_3 in Fig 4.13 harvested according to strategy 1? Explain your answer.

(c) What therefore are the dangers of using fixed quota harvesting when stock size is low?

(d) What could you recommend to fishermen fishing according to strategy 2 in Fig 4.17 that would increase their yield?

(e) What does the pink shaded area in Fig 4.18 represent?

4.10 (a) Compare and contrast the advantages of fixed quota and fixed effort harvesting.

(b) Why should fishermen, operating under a fixed effort regime, resist the temptation to fish for longer in a year when catches are poor?

(c) Why may it be impossible for fishermen to resist such a temptation?

4.11 Look at Fig 3.11.

(a) In 1913 which age class was the most important in this fishery?

(b) When did the fish in this age class recruit to the fishery?

(c) Account for the strength of this particular age class.

(d) What are the implications of the variation in age class strength which you can see in Fig 3.11 for (i) fisheries scientists setting quotas (ii) fishermen.

SUMMARY ASSIGNMENT

1. Explain the meaning of the following in relation to commercial fisheries: sustainable yield, surplus yield, maximum sustainable yield, fishing effort, catch per unit biomass, overfishing, underfishing.

2. What data do fisheries scientists need to collect to manage a fishery? Discuss how such data may be collected and the problems associated with its collection.

3. What is a surplus yield model? How is such a model constructed and how can it be used to manage a fishery?

4. Keep your answers to the analysis exercises and Questions 4.9, 4.10 and 4.11, as these will help you to remember how to read surplus yield curves and interpret surplus yield models.

Chapter 5

HARVESTING THE SEA 2: THE OVERFISHING PROBLEM

Despite all the best efforts of fisheries managers, all too often fisheries end up being overfished and destroyed. This is graphically demonstrated in the case of the northern Atlantic herring (Fig 5.1). In the 1950s some three thousand million herring a year were landed in British ports alone. However, the introduction of more powerful and bigger boats, combined with improved fishing technology, meant that by the early 1970s the once plentiful herring stocks had been destroyed by overfishing. Things were so bad that in 1976 the International Council for the Exploration of the Sea declared a ban on all catches of herring in the northeastern Atlantic, including the North Sea, until 1983. By 1984 stocks had recovered sufficiently for the European Community to lift the ban and fishermen were allowed to catch 84 300 tonnes of herring. However, the ban had effectively wiped out fishing ports like Lowestoft which relied on herring. In addition, the industry which processed and sold the fish virtually disappeared from Britain. Clearly, overfishing is disastrous biologically, economically and socially. In this chapter, we will examine how overfishing can be recognised, what advice fisheries managers can give to prevent overfishing and why such advice is often ignored. In addition, we will consider the importance of fish farming as an alternative to catching wild fish.

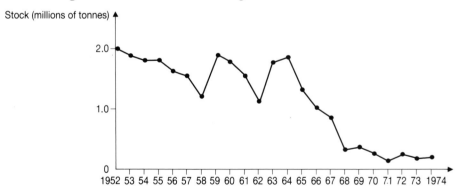

Fig 5.1 The decline of North Sea herring stocks as a result of overfishing.

LEARNING OBJECTIVES

After completing the work in this chapter you will be able to:

1. explain the causes and symptoms of overfishing;

2. outline the regulatory strategies which can be used to prevent overfishing;

3. provide examples of stocks which have been overfished and understand the complex origins of overfishing problems;

4. assess the present and future importance of aquaculture.

5.1 RECOGNISING AND PREVENTING OVERFISHING

Overfishing becomes apparent when fewer fish are being caught despite fishing effort being increased, and when the fish that are being caught become progressively smaller over time. There are two main types of overfishing.

1. **Recruitment overfishing** occurs when so many breeding adults are caught that insufficient eggs are laid to provide recruits which will replace their parents in the stock. This is a very serious state of affairs. For example, both the East Anglian and Atlanto-Scandian herring stocks collapsed through recruitment overfishing.

2. **Growth overfishing** occurs when too many small fish are caught, leaving fewer fish to grow to a large size. Since larger fish tend to sell for a better price this makes the fishery uneconomic, but is not of itself significant, providing sufficient fertilised eggs have been produced to maintain the stock.

Preventing overfishing

A whole battery of control measures can be introduced to prevent overfishing. These can be grouped into the main categories shown in Table 5.1. We will briefly discuss four strategies, but other strategies are discussed in the case studies in later sections.

Table 5.1 Fishery regulation measures and strategies used to achieve them.

Measure	Strategy
reduction of age at first capture	inspection of landed fish gear limitation, especially minimum mesh size

reduction of fishing mortality by
  restriction of fishing effort
	closed seasons/periods
	sanctuary zones
	gear limitation, e.g. banning
	  long drift nets
	catch quotas

enhancement and conservation of spawning grounds	preservation construction

Reducing the total permitted annual catch

This effectively means setting a lower quota. It not only causes political arguments between governments trying to set the quotas, but is also difficult to police. Thus, British fishermen in the North Sea haddock fishery, operating primarily from Peterhead, switch to catching whiting once they have filled their haddock quota. However, as similar gear is used to catch both species the fishermen inevitably catch haddock while fishing for whiting. These haddock, which are in excess of the quota, must be thrown back into the sea even though they will inevitably die. In addition, there is a great temptation to land only larger more marketable fish, throwing back the smaller fish which, even though they are dead, do not then count towards the boat's quota for that year. The discarding of dead and dying fish is called **slippage**. It is a particular problem with ultra efficient purse seiners, like those used in the west Scotland mackerel fishery, where a boat could catch its entire quota of fish in one night!

Regulating mesh size

This measure controls the age at which fish are first captured and helps to improve recruitment by allowing fish to grow sufficiently large to spawn.

Most fisheries set a lower limit on the size of the fish which can be caught by specifying a minimum mesh size, e.g. 90 mm in the North Sea cod fishery. However, on its own an increase in mesh size will not reduce fishing effort since there is no restriction placed on the number of boats entering the fishery, or on the length of time for which they fish. In addition, fishermen will naturally be resistant to introducing increased mesh sizes since it will inevitably lead to a reduction in their own personal catch. For example, introducing a mesh size of 120 mm into the North Sea haddock fishery would undoubtedly help to conserve fish stocks but it would, in the short term, drastically reduce the catches of fishermen.

Limiting the areas which may be fished

This provides sanctuary zones in which fish can grow to a reasonable size. It has worked well in the Pacific halibut fishery but requires investment in fishery protection vessels and staff. It may also be difficult to enforce when the boat fishing in a protected zone comes from a foreign country. Difficulty of enforcement can lead to fish wars, e.g. the cod wars between Iceland and Britain in the 1970s when the Icelanders extended their territorial limit to 200 miles and banned fishing within those limits by foreign boats (Fig 5.2).

Introducing closed seasons

This strategy also helps to reduce fishing effort as well as protecting fish at critical times in their life cycle, e.g. during spawning. Enforcement is again the major problem.

Fig 5.2 Fish are a valuable commodity which can be fought over. Here the Icelandic gunboat *Tyr* passes down the starboard side of a Royal Navy frigate which is protecting the British fishing fleet in disputed territorial waters around Iceland during the 'cod war' of 1976.

QUESTIONS

5.1 (a) Why would increasing the mesh size of nets reduce haddock catches in the short term but potentially increase them in the long term?
(b) Why do you think fishermen tend to think in the short rather than the long term?

5.2 Chesapeake Bay, on the east coast of the United States, supports an inshore **drift net fishery** for shad. In the late 1970s the shad **stock** was **overfished**, and was in danger of **recruitment failure**, and was having its estuarine spawning grounds damaged by dam building. The management plan for this fishery includes alternation of sanctuary areas, rest days when fishing is not allowed, restrictions on the size and mesh of the nets, and enhancement of the spawning grounds.
(a) Explain the meaning of the terms in bold.
(b) Suggest how the fisheries managers recognised that the stock was in danger of 'recruitment failure'.
(c) How will the management plan for the fishery help the stock to recover?

5.2 CASE STUDY: THE PERUVIAN ANCHOVETA INDUSTRY

This case study illustrates how mismanagement, political pressure and changes in the environment can lead to the collapse of a fishery. The anchoveta, *Engraulis ringens*, is a clupeid fish like the herring. It is a mainly herbivorous, planktonic feeder with a short generation time of about three years. (Compare this with plaice which may live up to 40 years.) The anchoveta flourishes in the upwelling which occurs in the currents off the eastern coast of Peru (see Fig 1.11). This brings cool nutrient-rich water to the surface and so provides ideal conditions for the growth of plankton. This process occurs predictably every year and provides the basis for the anchoveta industry.

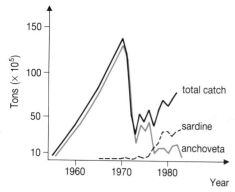

Fig 5.3 The collapse of the Peruvian anchoveta fishery. Note that fish catches are measured here in imperial tons not metric tonnes.

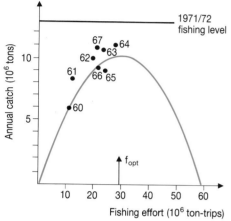

Fig 5.4 Surplus yield model fitted to the total catch of the Peruvian anchoveta. Note that the estimated catch from seabirds is included. Estimated MSY is about 10^7 tons in this model.

The development of the fishery

The fishery was initially based on drift netting using sail-powered boats. However, the seemingly limitless numbers of anchovetas encouraged the Peruvian government to encourage the expansion of the fishery throughout the 1950s and 1960s with a switch to purse seining. During the 1950s the fishery expanded at the rate of up to 174% per annum. As a consequence, by the early 1970s this one fishery provided 15% of the total world catch of fish and produced 65 million tonnes per annum. However, in 1972 the fishery collapsed (Fig 5.3). While an immediate moratorium on fishing would, perhaps, have encouraged the recovery of the fishery, by 1972 around 20 000 people relied on the anchoveta for their livelihood. In a desperately poor country like Peru it was politically unacceptable to ban fishing so it still continues today. Catches of the anchoveta remain low. So what went wrong?

Management of the fishery

The fishery was managed using a surplus yield model (Fig 5.4) which suggested a maximum sustainable yield (MSY) of around 10^7 tons and includes the estimated catch by seabirds. Fig 5.4 shows the level of fishing in 1971 and 1972. Clearly, the fishery was being overfished as a result of the large increase in fishing effort during the 1960s. This overfishing particularly affected the larger fish, with the result that, by October 1971, the fishery was suffering major recruitment problems; fish were being caught before they could spawn. However, the fishing continued despite this recruitment problem, but this on its own cannot explain the dramatic collapse seen in 1972. In addition, we need to take into account *El Nino*, a climatic event which affects the whole of the Pacific Ocean.

El Nino

Every 15 to 20 years the upwelling in the Peruvian current is disrupted or even stopped by *El Nino*. The cause of this major change in current patterns is still uncertain, but it seems to involve a change in the southeast trade winds blowing in the western Pacific. The effect, *El Nino*, causes warm, nutrient- poor, unproductive water to flow from the equator down the coast of Peru. *El Nino* was particularly pronounced in 1972 and led to a cessation of the upwelling in the Peruvian current. The resulting lack of food resulted in a high death rate among the larval anchovy. This compounded the loss of offspring produced by overfishing the adult spawners. In addition, *El Nino* caused the adult fish to congregate in a narrow band close to the shore, making them even easier to catch than usual.

The effects of *El Nino* were well known to the fishermen, if not to the fisheries managers. Similar *El Nino* events had occurred in the 1950s and 1960s. However, on these occasions the stock had both the age structure, as well as the reproductive capacity, to recover quickly. The overfishing of the early 1970s had destroyed this robustness and brought about an almost total failure in recruitment which resulted in the crash of the fishery. The unfortunate combination of a natural meteorological event and the over-confident application of too simplistic a fishery model, in combination with political pressure to expand the fishery to earn foreign currency, produced the crash.

The effects of the crash

The economic effects of the failure of this fishery were felt worldwide. For example, Peruvian fish meal provided a cheap source of protein rich in lysine, an essential component of animal feeds. Thus, as a result of the collapse of the anchoveta fishery, the British fish farming industry had to

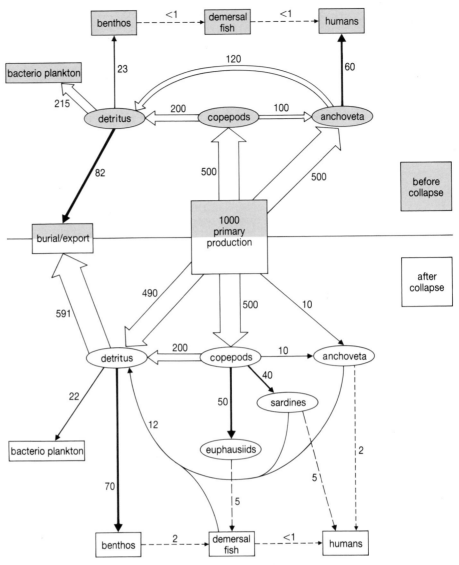

Fig 5.5 The food webs in the Peruvian current before and after the collapse of the anchoveta stocks in 1972.

switch to using food pellets made from pilchard meal from the South African pilchard fishery. Pilchard meal was twice as expensive as Peruvian fish meal; a potentially crippling increase in costs for an industry already beset by rising fuel costs as a result of the oil price rises of the early 1970s.

In addition, the collapse of the fishery has led to long-term, perhaps irreversible, changes in the food webs of the Peruvian currents (Fig 5.5). While primary production remains as high as before the *El Nino* of 1972 the food web has become much more complex. This means that less of the primary production is now going into fish that humans can use. The slight increase in the number of sardines caught since the collapse of the anchoveta stocks (see Fig 5.3) does not even come close to replacing the yield lost from this fishery. In addition to the human misery this has caused, up to five million seabirds, which also relied on the anchoveta, have been lost.

Such a pattern of overfishing combined with environmental changes has led to the collapse of several other fisheries based on clupeid fish, e.g. the northeastern Atlantic herring and the Californian anchovy fishery. Clupeid fisheries, in particular, seem to be difficult to manage.

5.3 (a) What harvesting strategy was being used to manage the Peruvian anchoveta industry?

(b) Why was this strategy unsuccessful?

(c) Did the fishery crash as a result of recruitment overfishing or growth overfishing?

(d) Fisheries managers did warn that the stock was in danger. Why do you think their warnings were not heeded?

(e) What alternative strategies could you have suggested to manage the fishery so that it provided a sustainable yield? What costs, economic and human, would your alternative strategies incur?

5.4 (a) Calculate, from the human point of view, the reduction in energetic efficiency of the food chains in the Peruvian current as a result of the collapse of the anchoveta fishery.

(b) What has happened to the primary production 'lost' to humans?

(c) Suggest possible reasons for the increase in the number of sardines caught, following the collapse of the anchoveta stock.

5.3 CASE STUDY: THE NORTH SEA FISHERIES

Fig 5.6 The fishing trawler *Antigua* beating up The Channel in a force 8 gale. Notice the small size of this side trawler which is typical of many of the boats which form the North Sea fishing fleet.

The North Sea is one of Europe's most important fishing grounds (Fig 5.6), providing fish for countries including Britain, Germany, Denmark, the Netherlands and non-European Community (EC) countries like Norway. Britain, for example, catches 40% of its cod, haddock and plaice in the North Sea and competition between countries and fishermen for access to North Sea fish stocks is fierce. As a result of overfishing both cod and haddock stocks in the North Sea are now perilously low.

The scale of the problem

According to data collected by the International Council for the Exploration of the Sea, spawning stocks of North Sea cod will be no more than 78 000 tonnes in 1991, compared with 168 000 in 1982. The spawning stock of haddock has fallen even more steeply, from 285 000 tonnes in 1982 to 81 000 tonnes in 1990. Since the mid-1970s the number of young haddock entering the seas around Britain has fallen from an estimated 122 billion to just seven billion.

Haddock and cod should live for more than ten years. They begin breeding when they are three to four years old. Now, fewer than one-third survive for more than a year after they have recruited to the adult stock. Even so, heavy fishing pressure on the stock can be offset by a good survival rate of the larval and juvenile fish. The survival rate was good in the early 1980s but has been poor since 1987 for reasons that are not really understood.

Fishery regulations in the North Sea

In 1977 the EC countries decided to adopt a 320 km limit around their waters. This was intended to stop non-EC countries from catching European fish stocks, and to allow member states to draw up plans to conserve stocks within the territorial limits.

In 1983 the EC drew up the common fisheries policy which is intended to run for 20 years. It set fishing quotas designed to limit the number of fish caught by individual member states. Political pressure during the negotiations to set these quotas undoubtedly resulted in the quotas being set too high. However, four successive years of cuts in the quotas, allocated to each of the EC member states under the common fisheries policy, have failed to prevent stocks from falling to their present, perilously low levels.

Even though early signals suggest that haddock numbers may recover in 1992, the outlook for cod remains poor. This led the International Council for the Exploration of the Sea's fishing advisory committee to recommend a 30% cut in the 1991 quotas for haddock, whiting and cod in addition to a 30% cut in 1990. This means, for example, that the 1989 British cod quota of 55 600 tonnes was reduced in 1990 to 46 180 tonnes; haddock quotas were 20 000 tonnes less in 1990 compared with 1989, with herring quotas being reduced by 15 000 tonnes in the same period.

In 1986 the European Commission also issued guidelines designed to persuade member countries to cut the size of their fishing fleets by 3% to reduce the fishing effort. However, the British fleet has increased to 9000 boats since 1986 and is now 20–30% too large. This massive increase in fishing effort has also coincided with new technological innovations which increase fishing efficiency.

Aids to better fishing

Since 1980 North Sea trawlers have become much more powerful and fishing gear has become more efficient. Consequently, catches have increased. Thus, while in 1975 a 27 m trawler would have had a 300 kW engine, the same size trawler in 1990 would have a 750 kW engine. This means that modern trawlers can use nets up to 100 m long, which has greatly increased the catches of bottom-living demersal fish such as cod, haddock, plaice and sole.

Modern equipment for locating fish includes fish-finders and sonars (Fig 5.7). A fish-finder sends an electronic signal down to the sea bed which bounces back giving information about the position of shoals of demersal fish like cod and haddock. Sonar uses a movable sound beam which can be used to search for shoals of pelagic fish like herring. It can tell the fishermen both the location and the depth at which the fish are swimming, allowing purse seine nets to be set accurately around the shoal. This enables the fishermen to trap entire shoals which, as the nets close up around the boats, can then be pumped on board.

The net problem

Trawl nets used in the North Sea characteristically have a mesh size of only 90 mm. These catch too many small fish: fish which cannot be sold because they are below the minimum landing size. In addition, the diamond-shaped mesh of these trawl nets (Fig 5.8(a)) closes up as the net is pulled through the water and this traps yet more small fish which have not yet had a chance to breed. These small fish are thrown back into the sea, although by the time that happens they are usually dead! Up to half the catch can consist of such discards.

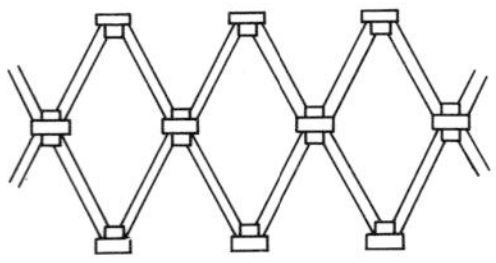

A further problem is that most of the North Sea fisheries are multi-species; in other words, the same techniques are used to catch different species of fish. This means a mesh size which catches only mature haddock also traps immature cod, which are bigger. One suggestion from the European Fisheries Commission was to increase the mesh size to 120 mm and to introduce a square mesh (Fig 5.8(b)) which does not close up like the diamond-shaped mesh. Trials conducted by the Sea Fish Industry Authority in Edinburgh and the Marine Laboratory in Aberdeen indicate that the 120 mm net would prevent the wasteful killing of small fish. However, the increased mesh size also allowed large numbers of mature fish to escape. For example, only eight haddock of marketable size were caught in the 120 mm net compared to 693 in the 90 mm net. In addition, fish like whiting, a species still in plentiful supply, slip through the 120 mm net.

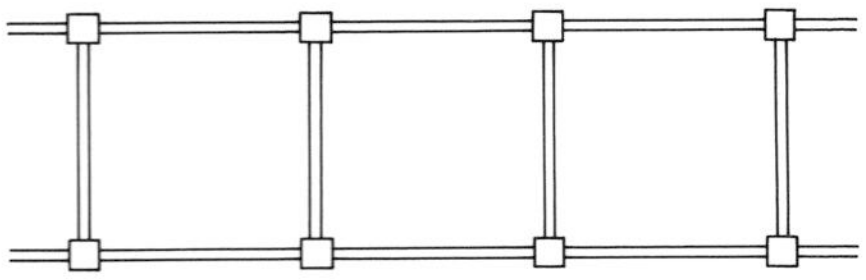

Further research, by the Sea Fish Industry Authority, suggests that the

Fig 5.7 Echo depth sounding equipment on a modern trawler. This sort of electronic sophistication makes the business of finding fish easier and so increases catches. The image shows both the depth of the water, 55.9 m, and a shoal of fish swimming at a depth of 20 m.

(a)

(b)

Fig 5.8 Different types of mesh. **(a)** Diamond mesh currently used in trawlers. This closes up preventing the small fish from escaping. **(b)** The square mesh that has been suggested as a replacement.

percentage of undersized fish caught in the standard 90 mm diamond mesh nets could be significantly reduced by including a square-mesh panel in the upper part of the net. Despite the determination of the then fisheries commissioner, Manuel Marin, an increase in mesh size was not included in the package of regulatory measures agreed in December 1990 by the European Fisheries Commission in an attempt to conserve dwindling stocks.

The human dimension

So far we have considered the biology and the economics of the fishing industry. To finish this section, we should also consider the human dimension. At the end of the day imposing restrictions on fishing, cutting quotas or changing mesh sizes mean affecting people's lives. The fishermen of Peterhead, who rely on cod, whiting and haddock for 60% of their catch of white fish, were encouraged in the early 1980s by government subsidies and grants to buy newer, more powerful boats. To do this they had to borrow large amounts of money. Initially, the smaller catches of the late 1980s were largely offset by higher prices; haddock was 42% more expensive in 1990 compared with 1989. However, the general downturn in the UK economy in 1990, leading to higher interest rates, coupled with fuel price rises as a result of the Gulf crisis, inevitably reduced fishermen's profits.

The fishermen know that stocks need to be conserved, but at the same time they have to feed their families and service their loans. However, the reality of the situation is grim. Fishing experts all agree that there are simply too many boats chasing too few fish. The obvious solution is to encourage fishermen to lay up their boats and so reduce the size of the fishing fleet. To this end the European Commission is recommending that fishermen be offered money to sell their boats and take up new professions. However, the British government is reluctant to implement this scheme, claiming that it would be too expensive and too difficult to administer. This leaves the fisherman faced with increasing costs, dwindling stocks and an understandable enmity towards further curbs on their activities. At the end of the day it is they who are caught between the devil and the deep blue sea.

QUESTION	5.5 (a) Summarise the reasons for the collapse of the North Sea cod and haddock stocks.
	(b) What regulatory measures already exist to conserve North Sea fish stocks? Explain how each of these regulatory measures works.
	(c) Outline a strategy which you think would help to conserve cod and haddock stocks further. What resistance do you think you would meet in implementing your plan? Would such resistance be justified?
	(d) Do you think fishermen should be paid not to fish? Explain your answer.

5.4 FISH FARMING

The principles of fish farming, or **aquaculture**, are no more than the common sense application of the biological principles that you have already learnt. A fish farmer aims to achieve the fastest rate of biomass increase possible, with due respect to cost, by:

• maximising increases in biomass due to growth and recruitment;

• minimising losses due to mortality.

To achieve this the chosen species needs to:

• grow well under captive conditions;

• be provided with appropriate food;

- convert food efficiently into fish flesh;
- be held under optimal conditions for rapid growth;
- be protected from predators, competitors and disease;
- (ideally) complete its life cycle under captive conditions.

Historical development

Aquaculture is not a modern invention. In the Far East, fish have been farmed for thousands of years while, in Europe, monks used to capture and keep carp in stock ponds for food. This type of subsistence or extensive fish farming, where the fish are primarily produced to provide food for the personal consumption of the farmer's family, is still common in developing tropical countries. Fish such as small carp, perhaps taken from a local river, are placed in a pond. The pond may be fertilised with animal manure which encourages the growth of the plants on which the carp feeds (Fig 5.9). Other than this, little else is done to increase growth or minimise mortality.

(a)

Fig 5.9 **(a)** Low-cost fish farming in Hong Kong. The carp in the pond feed both on the droppings from the ducks and on the algae whose growth is encouraged by the ducks' droppings. This pond therefore produces two valuable crops for the farmer's own consumption and for sale.
(b) High-cost fish farming in the United States. Huge numbers of salmon eggs are being incubated in the trays, under carefully controlled conditions, in this state-of-the-art hatchery.

Details of the slightly more developed extensive farming of milk fish are given in Fig 5.10. This herbivorous fish, related to carp, grows to a length of 100–180 cm and tastes like herring. It feeds in estuaries, lakes and rivers using gill rakers, but spawns in the sea. Such a system leads to a high rate of food production, and the utilisation of land which may be unsuitable for conventional farming, combined with low labour and food costs, ensures that this sort of fish farming is highly profitable (Table 5.2).

Table 5.2 Energy costs of some fish farming operations. Note that energy costs are a good indication of overall cost.

Species	Location	Type of farm	Energy costs (GJ/tonne harvested)
carp	Japan	intensive	300
milk fish	Philippines	intensive	5
milk fish	Philippines	extensive	1
trout	UK	intensive	55

Intensive fish farming

The primary aim of the fish farmers in a developed country is to make money, not to provide food for their own consumption. This has led to the development of intensive fish farming, of species like trout and salmon, where as many aspects as possible of the fishes life history are controlled.

The egg-bearing adults are spawned in captivity, or stripped by hand (Fig 5.11). The larvae are reared in tanks and the adults are grown to marketable size in special enclosures where they are fed on specially formulated diets (Table 5.3). Details of the intensive rearing of salmon are given in Fig 5.12.

Table 5.3 Food pellets used in raising rainbow trout.

Constituent	Proportion(%)
fish meal	26.3
dried milk	5
wheatmeal	30
salt, minerals, vitamins	23.7
meat and bone meal	15

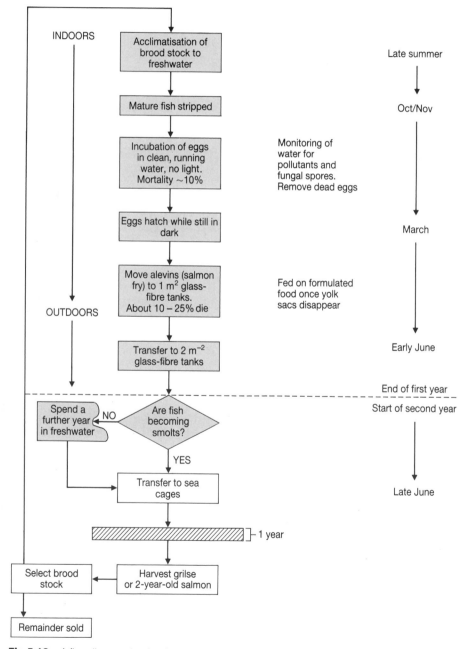

Fig 5.10 A flow diagram showing the stages in the farming of milk fish. Compare this extensive system with the intensive salmon-rearing system described in Fig 5.12.

Fig 5.11 Eggs being stripped from a live female rainbow trout. The eggs are then fertilised by sperm stripped from a selected male trout. Both fish are then returned to stock ponds.

Fig 5.12 A flow diagram showing the stages in the intensive farming of Atlantic salmon. Smolts are the small salmon that are ready to go to sea. Grilse are one-year-old salmon.

Marine fish farming

The most commonly eaten species of fish in Britain are all marine. Unfortunately, farming marine species is far more problematic than farming freshwater ones. Certainly, farming of marine molluscs, like mussels and oysters, has been practised for years and may provide high yields. Mussel culture on strings suspended below rafts off the coast of Spain produces 250 tonnes of flesh per hectare per year, while Japanese raft culture of oysters produces around 50 tonnes of flesh per hectare per year. However, shellfish, because they are sedentary, can be looked after relatively easily and can be protected from predators like starfish. Marine fish, however, move around a lot!

The fish could, of course, be corralled into enclosures made by damming a narrow inlet of the sea. However, this would interfere with the natural hydrography and may lead to potential problems with the temperature, salinity and oxygenation of the enclosed water. In addition, the enclosed water would need to be 'weeded' to remove potential predators and competitors before the fish were introduced. The fish would then need to be recaptured when they reached marketable size, which could be a major problem in a large piece of enclosed water. In addition, the marine species we like to eat are carnivorous so they sit at the end of food chains. This means that the transfer of energy to them, from primary producers, is less efficient than the transfer of energy to herbivorous species like milk fish. The only marine, herbivorous fish that we could eat that lives in British water is the grey mullet, a species which is not popular. The problem could be overcome by supplementary feeding, as in salmon-rearing pens (Fig 5.13) but this would be extremely expensive. Finally, the larval stages of marine species all require planktonic food of different sizes as they grow. This means that the farmer will need to raise large quantities of different types of planktonic food, something which again will add to the expense. Clearly, then, for farming of marine fish to be profitable the fish must fetch a very high price at the fish market.

There have been attempts to farm marine species. For example, salmon farming is successful, but it is only the adults of this species which live in salt water. An attempt was made to farm plaice and dover sole at two sites on the Scottish coast. The first was in an enclosed part of a sea loch at Ardtoe on the coast at Argyll. The second was in tanks at the Hunterston power station in the Firth of Clyde. The tanks were supplied with warm

Fig 5.13 High-protein food being fed to salmon in fish pens off the coast of Scotland. Excess food falling through the cage to the bottom of the loch forms a decomposing layer which severely alters the structure of natural benthic communities. The long term effects of this are as yet unknown.

sea water taken from the cooling towers of the power station. However, since plaice are a low-priced fish it became apparent in the late 1960s that farming this species was uneconomic. Farming the more expensive dover sole ran into problems because of the difficulty of feeding the fish in captivity. Another expensive fish, the turbot, has proved easier to rear in captivity. The larval stages feed initially on very small food items, supplied as rotifers, while the later stages can be raised on larger brine shrimps. The adults grow well in tanks, provided the temperature is closely controlled. They thrive at high densities, provide good food conversion efficiency and grow quite quickly, reaching 2 kg in two years under optimal conditions. The adults can be spawned in captivity so the whole life cycle can be controlled by the farmer.

Future advances

The critical activities in any fish farming operation are discussed below.

Continuous and cheap production of young fish

Further research is needed on the reproductive biology of fish species, particularly the hormonal control of spawning and improved fry handling techniques, to reduce mortality during the early stages of life.

The prevention of disease

Fish grown in high density in enclosures are particularly susceptible to diseases caused by five major groups of organism, ectoparasites, e.g. sea lice, fungi, endoparasites, bacteria and viruses. Prevention is best achieved by good hygiene and control by early detection, although more research is needed on controlling fish diseases. Treatment, where possible, takes one or two forms: dosing the water in the pond with a pesticide or treating the fish with drugs in their diet. Sulphonilamide antibiotics and nitrofurans in food are used to control many of the bacterial diseases affecting farmed fish. Sea lice, a particular problem in salmon-rearing pens in Scottish sea lochs, are killed by adding dichlorvos to the sea water.

Diet

Most of the research in this area is concerned with formulating dry diets which can be stored and transported in a dry pelleted form. Such a diet is easy to use and can be made to a standard formula using ingredients like 'trash' fish (small fish unfit for human consumption) or molluscs. However, such diets are expensive. For example, to raise 90 000–100 000 tonnes of Japanese yellowtail, a commercially valuable fish, requires the capture and processing of 500 000 tonnes of 'trash' fish.

Selective breeding

This aims to produce new strains of fish with faster growth rates, better food conversion efficiencies and resistance to disease. Methods used include inbreeding, heterosis, hybridisation and induced polyploidy. Details of these can be found in another book in this series, *Applied Genetics*.

Future developments

Currently, only 10% of the total aquatic foods we eat, including fish, shellfish and sea weeds, come from aquaculture (Table 5.4). Of this the overwhelming majority comes from freshwater species. Thus in Britain, apart from the still insignificant yield from turbot, practically all farmed fish are salmonids, particularly rainbow trout. While the FAO would like to see a five-fold increase in fish production from aquaculture over the next 20 years, MAFF (Ministry of Agriculture, Fisheries and Food) do not believe that aquaculture can make up for the shortfalls in the catches of marine fish.

Table 5.4 Estimated yields from world aquaculture.

Food	Yield (millions tonnes)
fish	4–5
sea weeds	1.1
oysters	0.6
mussels	0.2
other shellfish	0.15
total	6.0

Intensive farming of at least some species of fish does therefore appear feasible. However, the future development of the industry will rely largely on economic factors. In a nutshell, can sufficient fish of marketable size be produced at a price which ensures a profit for the farmer? Biological advances will of course help farmers to achieve their major aim of making a profit. However, the protein-rich food fed to the fish, which constitutes some 70% of the total cost of farming the fish, could be used in other ways. For example, if it were fed to chickens or pigs, which have faster growth rates than fish, it would result in the production of less expensive animal protein. For the foreseeable future then, intensive fish farming in developed countries should be viewed as a means of adding variety to our diets rather than as a source of cheap, plentiful protein.

QUESTIONS

5.6 Imagine a fish food could be made out of vegetable rather than animal protein. How and why would this improve the profitability of fish farming?

5.7 Why do you think diseases can sometimes spread so rapidly in a fish farm?

5.8 Ectoparasites of salmon include crustaceans called sea lice. The fish are treated with a pesticide dichlorvos applied to the water. Suggest the possible ecological impact of this pesticide on the native fauna near the rearing pens.

SUMMARY ASSIGNMENT

1. Make sure you have definitions and, where appropriate, examples of the following terms:
   overfishing, recruitment overfishing, growth overfishing, extensive fish farming and intensive fish farming.

2. Keep you answers to questions 5.3, 5.4 and 5.5 to remind you of why fisheries collapse and the consequences of overfishing.

3. Produce a plan for an essay to compare and contrast the objectives, processes and importance as protein sources of extensive and intensive fish farming. Use your plan to write your essay in about 40 minutes.

Theme 2

AGRICULTURAL ECOLOGY

In this theme, we are going to investigate the contribution that ecological research has made to agriculture. If you think about it the food you eat, the clothes you wear, even the paper you write on, are ultimately dependent on the primary production of plants. If we understand the ecology of this production process then we should be able to design agricultural methods and techniques which enable us to produce plants as efficiently as possible. Given that there are hundreds of new human mouths to feed, bodies to clothe and minds to educate every minute of every day then the importance of this aspect of applied ecology cannot be overemphasised.

Modern agriculture works on a vast scale using modern equipment to produce crops as cheaply as possible. Here a crop of oil seed rape, grown in enormous fields in Oxfordshire, is harvested using a combine harvester. The oil, extracted from the rape seed, is then used in a wide range of manufacturing industries.

Chapter 6

PLANT GROWTH ANALYSIS

A farmer's field is essentially a factory (Fig 6.1). Rather than producing, say, cars the plant factory produces organic molecules, carbohydrates, proteins, nucleic acids and so on. The art of good farming is to get the workers, the individual plants, to produce these chemicals as efficiently as possible. To achieve maximum efficiency the farmer needs to treat the fields in certain ways; ploughing them, providing drainage, fertilisers and so on. In addition, pests such as insects, fungi and viruses, which damage the plants and so impair their efficiency, and weeds, which compete with the crop plants for essential resources like water and light, need to be controlled.

The farmer needs to know when to plough, which crops to plant, how much fertiliser to use and when to spray pests, so ensuring that the land produces as high a yield as possible. This means that experiments need to be done which assess how different agricultural practices (**treatments**) affect plant performance. This raises a major challenge: how do you assess how well the plants are doing for a given treatment, say a fertiliser application? Plant growth analysis provides a quantitative and objective approach to answering this question. First though we need to look at the efficiency of photosynthesis since this will tell us how well the plants can do.

Fig 6.1 Maize growing in monoculture. By creating a wholly artificial environment, farmers ensure that each individual plant works at its maximum efficiency. Here, for example, the use of herbicides has completely eliminated weeds so that the maize plants derive maximum benefit from added fertilisers.

LEARNING OBJECTIVES

After completing the work in this chapter you will be able to:

1. evaluate the efficiency of photosynthesis in converting sunlight into chemical energy;

2. critically discuss the concept of growth;

3. define and use the following measurements of plant growth: absolute growth rate, relative growth rate, net assimilation rate (unit leaf rate);

4. define and use the following measurements of crop growth: leaf area index, crop growth rate, leaf area duration;

5. define and use the following measurements of crop quality: harvestable dry matter, harvest index, harvestable protein, digestible energy.

6.1 THE EFFICIENCY OF PHOTOSYNTHESIS

Some data on the efficiency of photosynthesis for a maize crop growing in the mid-west of the United States is given in Table 6.1. The results are rather surprising. Only 1.6% of the solar energy reaching the maize field during the crop's growing season was actually converted into carbohydrate by photosynthesis. Why is the process so inefficient? Energy contained in sunlight will be lost at many stages. Much of the sunlight will miss the leaves and fall on bare soil. The fate of the sunlight which does hit the leaves is shown in Fig 6.2.

Table 6.1 Energy budget of one acre (4047 m²) of maize during one growing season of 100 days.

	Glucose (kg)	Energy (10^6 kJ)	Percentage of incident solar energy
incident solar energy	—	8580	100
NPP of maize	6687	105.6	1.2
respiration (R) of maize	2045	32.3	0.4
GPP (NPP + R)	8732	137.9	1.6

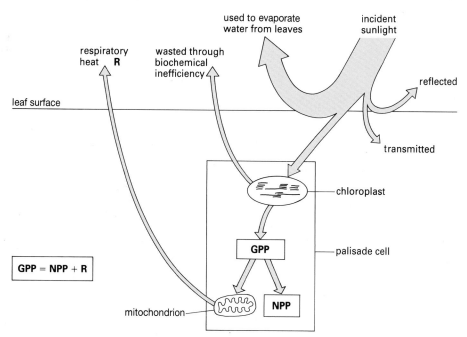

Fig 6.2 The fate of sunlight hitting a leaf. Notice how little actually reaches the chloroplasts.

A large proportion of the sunlight which strikes a leaf is not used in photosynthesis. Thus ultraviolet (wavelength < 380 nm) and infrared (wavelength > 720 nm) light, which constitute some 56% of the energy of solar radiation at ground level, cannot be used at all in photosynthesis. These wavelengths are either transmitted or reflected or, when absorbed, do not excite the electrons in the photosynthetic centres of photosystem I or II but merely raise the temperature of the leaf. The fate of the remaining 44% of the incident light energy, lying between 380 and 720 nm, called **photosynthetically active radiation (PAR)**, also depends on its wavelength (Fig 6.3). Only 85% of the PAR which strikes a leaf is actually absorbed by the plant's photosynthetic pigments.

A surprising amount of the energy of sunlight which is actually trapped by chlorophyll is wasted as a result of biochemical inefficiencies. Theoretically, the maximum photosynthetic efficiency is about 8%. In practice, the maximum rates for crop plants are of the order of 3–4% with a mean value of about 1%. For example, sugar beet growing in SE England has a maximum recorded photosynthetic efficiency, under experimental conditions, of 4% with a mean efficiency throughout the growing season of 1.2%. Equivalent figures for sugar cane growing in Hawaii are 3% and 0.8%. From this value you also need to subtract the energy inputs involved in mechanised, intensive agriculture; diesel fuel for tractors, the energy needed to make fertilisers, pesticides and so on. Clearly, agriculture has to work with a process, photosynthesis, which is not very efficient. The farm-

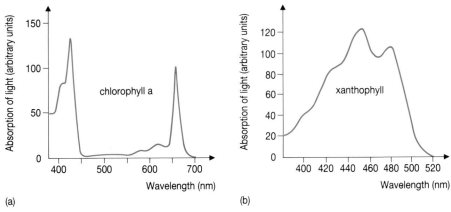

Fig 6.3 **(a)** The absorption spectrum of chlorophyll. Notice how it is only the red and blue wavelengths of light that are in the main absorbed. Green wavelengths tend to be transmitted.
(b) Accessory pigments, like xanthophyll, extend the range of wavelengths of light which can be used in photosynthesis.

ers job is to try, by adopting appropriate crop management procedures, to keep that efficiency as high as possible and so maximise crop yields.

QUESTIONS

6.1 Why does chlorophyll appear green?

6.2 **(a)** Why can raising animals be considered an inefficient way of providing food for human consumption?
    (b) Under what circumstances will raising animals be the only way of providing food for human consumption?

6.2 MEASURING PLANT PERFORMANCE

One way of measuring plant performance would be to measure the rate of photosynthesis, i.e. how fast the plants are producing carbohydrates, under a range of different treatments. However, a plant's performance is more than just the production of carbohydrates during photosynthesis. We need to know the proportion of these newly synthesised carbohydrates that are turned into proteins and nucleic acids. How efficiently are the plants using the available sunlight and water? What effects does a lack of nutrients have on plant performance? We could answer all these questions by carrying out experiments which investigated just one specific aspect of plant performance but what we really need is one simple measure which somehow integrates all these physiological aspects of plant performance. The answer is to measure **plant growth**.

What is growth?

If we are going to use growth as a measure of plant performance then it is essential that we should have a clear understanding of what we actually mean by the term and how we can measure it. Just think about yourself for a moment. You started off life as a single cell. You now have a body which consists of many different sorts of cell. You have obviously increased in size (both volume and mass) and complexity. So it seems that for you at least growth has involved:

• an increase in cell number;

• an increase in mass;

• an increase in volume;

• an increase in complexity.

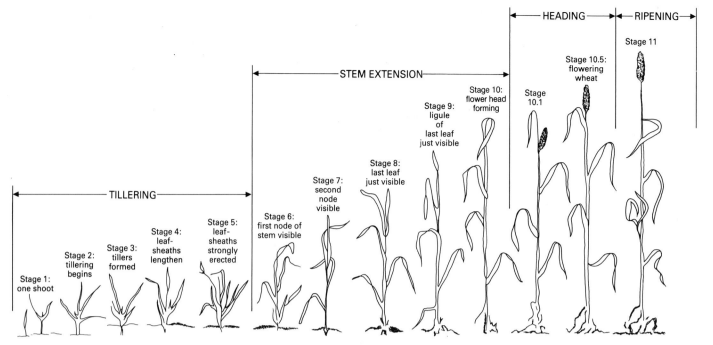

Fig 6.4 Stages in the growth of cereals. Tillering is the development of side shoots.

The same sorts of change occur as a plant grows. It starts life as a few cells in the embryo of a seed. The seed germinates and the resulting plant gets bigger, contains more cells and becomes more complex, developing roots, flowers and so on (Fig 6.4).

Clearly, then, there is no difficulty in producing an **intuitive definition** of what growth is but there is a problem of providing an **operational definition** of growth. Such a definition must involve some idea about the way in which you measure growth in a quantitative manner.

Measuring growth

What we are looking for is some simple measure which somehow encapsulates what growth is. We could measure the increase in cell number and volume but this is not easy and certainly could not be applied to measuring the growth of large numbers of plant. We could devise some way of measuring the increase in complexity as a plant grows but plants grow in so many different ways that such a measure would inevitably be rather vague and arbitrary. So that leaves us with measuring changes in size. We could measure changes in volume but this presents two problems.

1. It is difficult. Can you devise a simple, rapid method for measuring the volume of, say, a wheat plant (see Fig 6.4)?
2. The volume of a plant changes depending on how much water it contains.

So that leaves us with measuring changes in mass.

Normally, increases in dry mass are used as a measure of plant growth but even here this presents problems. Look at Fig 6.5. This shows the change in dry mass of maize through the growing season plotted on a logarithmic scale. Note the following points.

- There is no change in dry mass for about the first ten days of growth.
- The plant then loses mass until almost 20 days have passed. During this time there are a number of changes in the young seedling – development of leaves and roots for example.

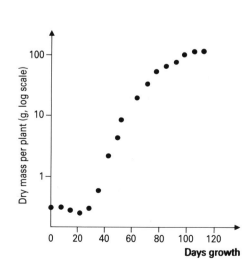

Fig 6.5 Changes in the dry mass of maize. Notice the logarithmic scale which enables you to see the small decline in mass which occurs early on in the growth of this plant.

- After 20 days the leaves begin to produce carbon compounds in excess of the respiratory needs of the plant and there now follows a period of rapid growth during which both the dry mass and the complexity of the plant increase.
- Maize flowers at about 65 days after germination. After this time more and more of the plant's resources are directed into developing seeds on the cob.
- After about 105 days there is no further increase in dry mass of the whole plant but the cob is still increasing in size, i.e. it is still growing although by the criterion of increase in dry mass, plant growth has stopped.

Clearly, then, no firm definition of growth can be given. An **irreversible increase in dry mass** is a reasonable working definition of growth in individual plants but even this presents problems when we look at the growth of a plant throughout its complete life cycle. Often, as we will see, ecologists are more interested in measuring changes in form or leaf area rather than changes in dry mass.

Measuring growth rates

Growth rate refers to an increase in plant size per unit time. The growth of plants can be measured in two different ways.

1. **Instantaneous measurements** tell us the growth rate of a plant now.
2. **Average measurements** tell us the growth of a plant over a period of time, say a week. Such data is then often converted to daily growth rates. Thus, if a plant increases in mass by 7 g over the period of a week then its daily growth rate is 1 g day^{-1}.

Working out instantaneous measurements is time consuming and involves quite complex mathematics, usually performed by computers. While this approach is now gaining ground most agricultural ecologists still measure average growth rates, and we will use these in this book. However, there is a danger in using average measurements.

You can appreciate this problem if you imagine we are trying to measure the speed of a motor car. If I drive from my house into Manchester, a distance of about 40 km, I can measure my speed in one of two ways. I can look at the speedometer which gives me an instantaneous measurement of my speed or I can time how long it takes me to cover the 40 km, which will give me an average .speed. However, if we consider the actual (instantaneous) speeds I drove at on my 40 minute journey we will find that there is a lot of variation in the actual speed of my car. Similarly, an average growth rate may cover up a lot of variation in the instantaneous growth rate of the plants. Such variation will be particularly large when plants are growing rapidly. We can partly overcome this problem by ensuring that we measure average growth rates over short periods of time, particularly when plants are growing rapidly.

Collecting growth data

Having decided to measure average growth rates we now need to collect **primary data**. This involves harvesting plants and measuring:

- total dry mass per plant or area of crop (*M*);
- leaf area of plant (*LA*);
- dry leaf mass per plant (*LM*) or area of crop.

This data will be collected at several points during the course of the growth of the crop. Since the sampling technique is destructive – the dry mass of a plant can only be measured once – we will need to take large, representative samples of the crop at each harvest. The primary data collected at each harvest can then be used to calculate plant growth using one of the methods described in the next section.

QUESTION

6.3 (a) Why should the interval between harvests be made shorter when plants are growing rapidly or when they are undergoing rapid changes in structure and physiology?

(b) Why is a large sample of plants necessary when calculating average growth rates?

6.3 GROWTH ANALYSIS OF INDIVIDUAL PLANTS

In this and the subsequent section a series of methods used in measuring plant growth are outlined. To ease comparison each description has the same format: a definition, the determination of the measurement, some examples and a discussion of the technique.

Absolute growth rate

Definition: The absolute growth rate (G) is the average increase in plant dry mass per unit time.

Calculation:

$$G = \frac{M_2 - M_1}{T_2 - T_1} \tag{1}$$

where M_2 is plant dry mass at time T_2 and M_1 is plant dry mass at time T_1. To illustrate the procedure look at the data in Table 6.2 for radish plants grown in a greenhouse. To calculate G using equation (1): $T_1 = 70$ days, $M_1 = 7.16$ g; $T_2 = 77$ days, $M_2 = 7.65$ g; and so:

$$G = \frac{7.65 - 7.16}{77 - 70} = 0.07 \text{ g day}^{-1} \tag{2}$$

Table 6.2   The dry mass of radish plants 10 and 11 weeks after planting.

Time (days)	Plant dry mass (g)
70	7.16
77	7.65

Discussion: The problem with absolute growth rate is that we cannot use it to compare the growth of plants of different size. Imagine two plants which have been grown for a week in an experiment. Plant A weighs 1 g at the beginning of the experiment while plant B weighs 10 g. At the end of a week both plants have increased their mass by 1 g. Their absolute growth rate is the same. But does this really reflect how well the plants are performing: how fast they are growing? After all, plant A has doubled in size but plant B has only increased its mass by 1/10. Clearly, then, we need a measure of plant growth which takes account of differences in plant size.

Relative growth rate

Definition: The relative growth rate (RGR) is the increase in plant mass per unit mass per unit time.

Calculation:

$$\text{RGR} = \frac{\log_e M_2 - \log_e M_1}{T_2 - T_1} \tag{3}$$

The use of natural logarithms in this and other equations in this chapter arises from the use of calculus. You do *not* need to understand calculus to use the equation.

where $\log_e M_1$ is the natural logarithm of dry mass at time T_1 and $\log_e M_2$ is the natural logarithm of dry mass at time T_2. Your calculator will probably be able to give you the natural logarithm of a number. We can use the example of plant A and B given above to illustrate the calculation, which is set out in Table 6.3.

Discussion: RGR provides an index which allows us to compare the growth of different species or the same species growing in different environments or under different conditions. However, a major drawback in

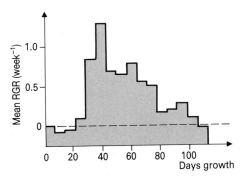

Fig 6.6 Changes in mean relative growth rate with time for the maize crop shown in Fig 6.5.

Table 6.3 Calculating relative growth rate (RGR)

Quantity	Plant A	Plant B
M_1	1 g	10 g
M_2	2 g	11 g
$\log_e M_2 - \log_e M_1$	0.693–0	2.40–2.30
$T_2 - T_1$	7 days	7 days
RGR (g day^{-1})	0.099	0.014

the use of RGR is the amount of variation the index shows during the growing period of the plant (Fig 6.6). Some of this variation may be due to changes in the environment during the growing season, changes in light intensity, the availability of water, nutrients and so on. However, many of the changes in RGR seen in Fig 6.6 are due to changes in the plants themselves as they grow, so-called **ontogenetic drift**. Such changes include:

- changes in the overall structure and proportions of the plant, for example changes in leaf area relative to overall plant biomass;
- alterations in the fate of photosynthetic products, for example being used to produce fruits rather than new leaves;
- changes in plant metabolism as a result of ageing.

In practice, it is difficult, indeed almost impossible, to disentangle the effect of changing environmental conditions and ontogenetic drift on the variation in RGR. The upshot of this is that meaningful comparisons of plant performance between different treatments, say different rates of fertiliser application, is difficult using RGR.

In addition, the concept of RGR implies that all that is necessary to produce more plant mass is plant mass. This is simply not true. As plants grow the proportion of dead, structural support material increases. This 'necromass' has no role in the production of new plant biomass and yet it is taken into account when RGR is calculated. We can get around this problem by using an index of plant growth which relates the increase in plant biomass to the area of leaves, the plant organs responsible for producing most of the organic molecules needed for growth.

Net assimilation rate or unit leaf rate

Definition: The net assimilation rate (NAR) or unit leaf rate (ULR) is the net increase in plant biomass per unit leaf area per unit time.

Calculation: This is a little more complex than the previous two methods since we need to make some assumptions about the relationship between plant biomass and leaf area. If there is a linear, (straight line) relationship then the following equation can be used:

$$\text{NAR} = \left(\frac{M_2 - M_1}{T_2 - T_1} \right) \left(\frac{\log_e LA_2 - \log_e LA_1}{LA_2 - LA_1} \right) \tag{4}$$

where LA_1 is the leaf area at time 1 and LA_2 is the leaf area at time 2, with all other symbols as before. A method for determining leaf areas is given on page 81. However, the assumption of linearity may not be valid if:

- the plant is growing rapidly;
- the intervals between harvests ($T_2 - T_1$) is too great.

Under these circumstances other methods have to be used.

Discussion: NAR, like RGR, is an index of growth that is independent of plant size. While it is more reliable as a measure of plant growth than NAR it still exhibits some ontogenetic drift. Nonetheless, NAR does provide the basis for a sounder comparison of individual plant performance under different treatments than does RGR. In particular, the environmental influences on NAR are much more than internal plant processes. Thus the plants in Fig 6.7 all reach their maximum NAR at approximately the same time despite being sown at different times and therefore being at different stages of development, i.e. there is little effect of ontogenetic drift on NAR.

QUESTIONS

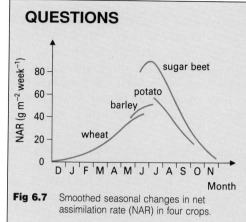

Fig 6.7 Smoothed seasonal changes in net assimilation rate (NAR) in four crops.

6.4 (a) Calculate the RGR of the radish plants from the data given in Table 6.2.
(b) Predict what will happen to RGR as the proportion of leaf area to total plant biomass increases.
(c) Why will NAR be less affected than RGR by changes taking place in the ratio of leaf area to total plant biomass?

6.5 Why do you think the NAR of all four crops shown in Fig 6.7 peaks in June or July?

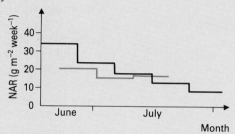

Fig 6.8 Changes in NAR in barley plants grown in two environments: outdoors (black line) and in growth chambers (red line).

6.6 Fig 6.8 shows the results of an experiment on barley where plants were grown in either a controlled environment indoors or outside in a field.
(a) Explain the decrease in NAR with time in the growth room.
(b) Explain the greater decrease in NAR of the crop grown outside.
(c) Why does this experiment support the view that NAR is more sensitive to changes in the environment rather than ontogenetic drift?

6.4 GROWTH ANALYSIS OF POPULATIONS AND COMMUNITIES OF PLANTS

All the measures that we looked at in the previous section were concerned with the performance of individual plants: the results that we obtain from the calculations are expressed on a per plant basis. However, farmers and agricultural ecologists are not usually concerned with the performance of individual plants but rather with the performance of whole populations of plants (e.g. cereals growing in monoculture) or communities of plants (e.g. the mixture of grass species growing in a pasture). In theory, we could use RGR or NAR to measure plant performance on this basis but, in practice, another set of plant growth analysis techniques are used. This section is concerned with these techniques. Each method will be presented in the same format as that used in Section 6.3: definition followed by method of calculation and discussion.

Leaf area index

Definition: The leaf area index (LAI) is the total projected leaf area per unit area of land.
Calculation:

$$LAI = \frac{L}{A} \qquad (5)$$

where L is the total projected leaf area above a land area A. Since the units for L and A are both the same (i.e. of area) LAI is a dimensionless ratio. For example, if $L = 2.7$ m^2 above a land area of 0.5 m^2 then LAI = 5.4.

To measure each of these parameters you would harvest all the leaves above a known area of land, say 0.5 m^2 = A. You then need to determine the area of one side, the **projected area**, of all the leaves you have harvested which gives you L.

Leaf area can be determined by punching discs out of a known mass of leaves using a cork borer of known diameter. Count the number of leaf discs and weigh them. Since you know the diameter of the cork borer you can calculate the area of each leaf disc. Multiplying this by the number of discs will give you the total leaf area you have punched out. The mass of the leaf discs will now enable you to convert leaf area into leaf mass. Dividing this number into the total mass of the leaves will give you the leaf area. Dividing by two will give you the projected leaf area.

Discussion: The concept of LAI is an attempt to reduce the growth of a crop to the growth of a population of leaves displayed by the crop – a measure of 'crop leafiness' if you like. Effectively, LAI is a measure of the number of complete layers of leaves displayed by the crop. The more layers the higher the LAI (Fig 6.9). This is a major limitation of the LAI concept since leaves never actually form unbroken layers lying, one above the other. Instead, leaves are held by plants at varying angles to the horizontal (Fig 6.10) and these angles vary with the plant species or variety, the conditions under which the plant is grown and the age of the plant. Furthermore, LAI also assumes that the leaves at all layers in the canopy are functionally the same. This is not the case. For example, leaves lower in the canopy may actually respire faster than they can photosynthesise because they are shaded by leaves higher in the canopy (Fig 6.10). Such leaves are actually a drain on the plant's resources compared with leaves higher in the canopy. Despite these limitations LAI has proved a useful and popular index of crop performance.

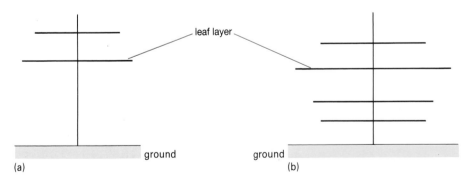

Fig 6.9 Leaf area index (LAI) depends on the number of layers of leaves in a plant's canopy. Here plant (b) will have a greater LAI than plant (a).

LAI varies with the stage of crop development. In a newly germinated crop LAI will be less than 1 since the leaf area of the young seedlings will be tiny in relation to the area of land sampled (Fig 6.11). As the crop grows it produces more and more leaves which increase in size until LAI reaches a maximum value, usually between two and ten for crops grown in the British Isles and other temperate areas.

Changes in LAI are more independent of seasonal changes in the environment than is NAR. This is clearly illustrated by comparing Figs 6.7

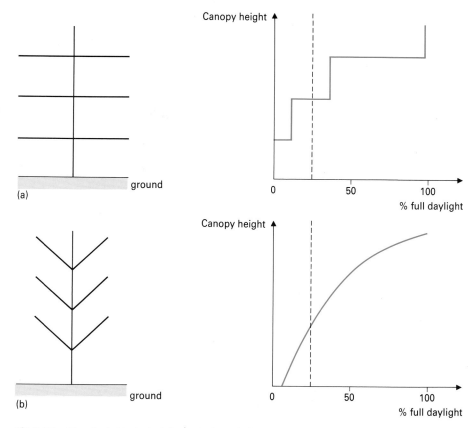

Fig 6.10 The effect of leaf orientation on leaf area index and light interception. Crop (b), e.g. wheat, will have a higher LAI than crop (a). Note also the sharp reduction in light intensity at the top of the canopy in (a) compared with a more gradual change in light intensity through the canopy of crop (b). This means that leaves lower in the canopy of (b) will be above the compensation light intensity (the vertical dotted line).

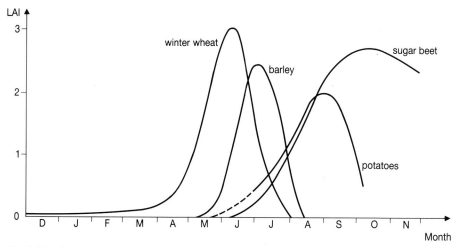

Fig 6.11 Smoothed seasonal changes in leaf area index (LAI) in four crops. Compare this diagram with Fig 6.7 and note that the time of maximum LAI depends on the time of planting and not the time of maximum incident light.

and 6.11. In all four species NAR is at its highest in the middle of the year. By contrast, changes in LAI are more strongly affected by sowing date than by seasonal changes in the environment. Thus frost-resistant winter wheat sown in the previous autumn reaches maximum LAI in late May/early June, which is some three months earlier than a potato crop which has to be sown in late spring because potatoes cannot tolerate frost damage.

Overall crop performance will be controlled by two factors.

1. The efficiency with which leaves produce new biomass, measured by NAR.
2. The leafiness of the crop, measured by LAI.

It would be useful if we could combine these two values to give a single index of agricultural productivity. This is the idea behind crop growth rate.

Crop growth rate

Definition: The crop growth rate (CGR) is the increase in crop mass per unit area per unit time.

Calculation:

$$CGR = \frac{1}{A}\left(\frac{M_2 - M_1}{T_2 - T_1}\right) \tag{6}$$

where M_1 and M_2 are the dry masses of the crop harvested from equal but obviously separate areas of ground of area A at time T_1 and T_2. If the masses are expressed per unit area then:

$$CGR = \left(\frac{M_2 - M_1}{T_2 - T_1}\right) \tag{7}$$

Mean crop growth rate expressed in this form is an absolute growth rate – a difference in mass divided by a difference in time.

Discussion: CGR, at any instant in time, is the product of LAI and NAR (i.e. CGR = LAI × NAR). The direction and change in CGR will therefore mirror changes in NAR and LAI as shown in Fig 6.12.

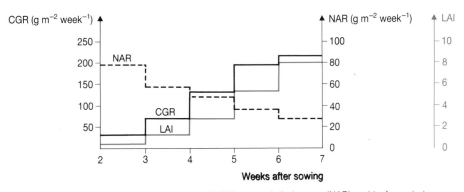

Fig 6.12   Changes in mean crop growth rate (CGR), net assimilation rate (NAR) and leaf area index (LAI) in relation to the age of wheat. Note that it is LAI which is more important in determining CGR than NAR.

CGR provides a useful means of comparing the performance of different crops and the effects of different treatments on crop performance. For example, daffodils planted at high densities (Fig 6.13(a)) and attaining high LAI had increased CGR despite a reduction in NAR. Conversely, removal of leaves of kale, so reducing LAI to a value of 3–4, actually led to an increase in NAR and CGR, suggesting that kale yields might benefit from the plants being well spaced and thinning of the crop (see Fig 6.13(b)). Sugar beet yields appear to be greatest at higher values of LAI, between 6 and 9, suggesting that yields of this crop will benefit from closer planting.

Leaf area duration

Definition: This is best given by looking at Fig 6.14 where LAI has been plotted against time. The area under this curve is the leaf area duration

Fig 6.13   (a) The plants which constitute this flower crop can be planted close together because they have an erect growth habit. **(b)** Young forage kale plants. Notice how the leaves are beginning to adopt a more horizontal growth position. As this crop develops the leaves will shade the ground to such an extent that the current, rampant weed growth will be suppressed without the use of herbicides. Notice the large distance between the individual crop-plants

PLANT GROWTH ANALYSIS

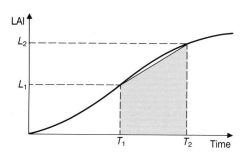

Fig 6.14 Leaf area duration can be determined by plotting the leaf area index against time. It is shown here between time T_1 and T_2 by the red area.

(LAD). This is a measure of the total opportunity that a crop has for producing organic material during the period in question. Units in this case will be those of time since LAI is dimensionless.

Calculation: The method chosen will depend on the mathematical skill and knowledge of the operator.

1. If the equation for the curve in Fig 6.14 is known then LAD can be determined by integrating the equation between the two chosen time intervals.
2. The curve can be traced on to paper and the area of interest under the curve cut out. This area can then be determined.
3. The shaded area under the curve in Fig 6.14 approximates to the shape of a trapezium. The area of this trapezium is an estimate of LAD between the times shown and can be calculated as:

$$\text{LAD}_{T_1-T_2} = \frac{(LA_1 + LA_2)\,(T_2 - T_1)}{2} \qquad (8)$$

You could actually estimate the whole of the area under this curve by dividing it up into a series of trapeziums such that the top edge of each trapezium approximates closely to the shape of the curve. If you then add up all the areas of all the trapeziums that will give you an estimate of the area under the curve.

Discussion: Combining data on LAD and NAR shown in Table 6.4 provides the opportunity for estimating crop yield since:

$$\text{Yield} = \text{LAD} \times \text{NAR} \qquad (9)$$

Note that the relationship is only approximate since mean NAR varies so much during a growing season. Of the two factors, LAD has been shown to be more important than NAR in determining yield.

Table 6.4 A comparison of yield, leaf area duration (LAD) and net assimilation rate (NAR) in four crops. Mean NAR is derived by dividing yield by LAD. The fairly constant number which results means that LAD is a reasonably good predictor of final yield.

Crop	Yield (tonne ha^{-1})	LAD (weeks)	Mean NAR (tonne ha^{-1} week^{-1})
barley	7.3	17	0.43
potato	7.7	21	0.36
wheat	9.5	25	0.38
sugar beet	12.0	33	0.36

ANALYSIS

Putting all the measures together

This exercise will help you to see how all the various measures of plant growth that we have looked at so far fit together and will give you practice in answering exam questions.

Fig 6.15 shows the average changes with time in leaf area index and net assimilation rate of winter wheat (W), barley (B), potatoes (P) and sugar beet (S).

(a) Define the term 'net assimilation rate'.

(b) In order to calculate net assimilation rate it is necessary to measure leaf area and dry mass. Explain how these values are measured.

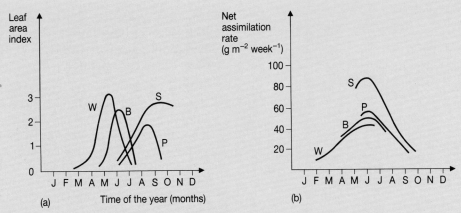

Fig 6.15 Changes in the LAI and NAR of four crops with time.

(c) (i) With reference to Fig 6.15(a), explain the rise and fall of leaf area index with time. (ii) Suggest why the peaks of leaf area index for different crops are at different times of year.

(d) Explain the influence of leaf area index on crop productivity.

(e) Sugar beet and potatoes are both planted in mid-April. (i) What evidence is provided by the graphs that earlier planting of these crops would produce higher yields? (ii) Suggest one reason why farmers do not plant these crops earlier?

(f) Using information in both graphs, determine the approximate crop growth rate of barley in early July. Show how you arrive at your answer.

(g) Discuss the environmental, structural and physiological factors responsible for the changes in net assimilation rate in Fig 6.15(b).

QUESTIONS

6.7 (a) Explain why if LAI exceeds the optimum value the performance of the crop may actually fall.

(b) NAR usually falls as LAI increases. Why?

(c) Why will the more erect foliage of sugar beet support a higher NAR for high values of LAI than the spreading leaves of crop like kale?

(d) Why does LAD × NAR provide an approximate measure of plant yield?

6.8 Account for the high CGR of daffodils compared with kale.

6.5 MONITORING CROP QUALITY

The indices discussed in the previous two sections have been concerned with measuring the increase in total biomass of a plant or crop. However, managing a crop on this basis may be meaningless when a farmer is only interested in a specific part of a crop, say grain, or where the concern is the quality rather than the quantity of the crop. Under these circumstances, we need a new set of indices on which to base a programme of crop management.

Harvestable dry matter

Definition: The harvestable dry matter is the amount or proportion of dry matter capable of being harvested from a crop.

Calculation: This will depend on what you mean by the 'amount of dry matter capable of being harvested'. Consider a farmer making hay or silage

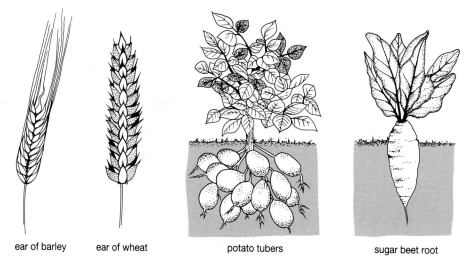

ear of barley   ear of wheat   potato tubers   sugar beet root

Fig 6.16 Only parts of these crops are economically valuable.

for winter cattle feed. All the grass harvested, the **biological yield**, is economically useful, as the cattle can eat all of it. So here the **economic yield** from the crop is the same as the biological yield. But now consider a cereal crop like wheat where we are only interested in the grain. This time the biological yield, all the plant biomass produced during the growing season, is not economically useful. The farmer is only interested in a small proportion of the total biological yield, the grain (Fig 6.16). So here the economic yield is only a small fraction of the biological yield. This proportion of the biological yield which represents economic yield is called the **harvest index**, i.e.

$$\text{Economic yield} = \text{biological yield} \times \text{harvest index} \qquad \textbf{(10)}$$

Successful management of crops like wheat does not depend solely on raising the biological yield but in maximising the harvest index. Thus over-application of fertilisers may increase the leafiness of a cereal crop, increasing its LAI, but it reduces the amount of light which can penetrate the canopy and inhibits seed set, so lowering the harvest index.

Harvestable protein and digestible energy

Farmers often have to take into account not only the quantity of a particular product that they produce but also its quality. This is particularly important if the crop is being grown for a particular market, e.g. malting barley for brewing, wheat for bread making or animal feedstuffs. Here both the amount of protein, the harvestable protein, that the crop contains and its digestibility are important. Digestible energy is defined as the amount or proportion of the energy content of a plant which is capable of being digested or absorbed by an animal.

QUESTION	6.9 Short-stemmed rice varieties have harvest indices of around 0.54 while taller, leafier varieties have harvest indices of around 0.41.

    (a) Which of the two varieties will have a higher economic yield given the same biological yield? Account for your answer.

    (b) Account for the increase in the harvest index of the wheat varieties listed in Table 6.5.

    (c) Why might a farmer grow a variety of wheat which does not necessarily have the highest harvest index?

Table 6.5 Harvest index, stem length from soil to base of ear and annual yield for a range of varieties of winter wheat (*Triticum aestivum*).

Variety	Year introduced	Harvest index	Stem length (cm)	Grain yield (tonnes ha$^{-1)}$)
Little Joss	1908	0.36	1420	5.22
Holdfast*	1935	0.36	1260	4.96
Cappelle Desprez	1953	0.42	1100	5.86
Maris Widgeon*	1964	0.39	1270	5.68
Maris Huntsman	1972	0.46	1060	6.54
Hobbit	1977	0.48	800	7.30
Norman	1980	0.51	840	7.57

* These are 'bread-making' wheats which have a slightly lower yield than the contemporary 'feed' wheats used in making animal feeds.

SUMMARY ASSIGNMENT

1. What are the advantages and disadvantages of using plant growth as a measure of plant performance?
2. Ensure you have definitions and appropriate examples for all of the terms given in the learning objectives.
3. Design and preferably carry out an experiment(s) which would enable you to calculate all of the measures of plant growth given in this chapter for mustard, a fast growing plant.
4. Keep your answers to *Analysis: Putting all the measures together* to remind you how measures of plant performance can be used and are related to each other.

Chapter 7

CLIMATIC FACTORS AFFECTING PRODUCTIVITY

Fig 7.1 summarises the problems faced by a food producer trying to turn wheat into bread. The objective is to turn as much of the gross primary production (GPP) into bread as possible. The efficiency of this process will depend on:

- providing a sufficient supply of essential resources – sunlight, water, nutrients and carbon dioxide – so that the processes of photosynthesis and biosynthesis can proceed at their maximum rate;

- managing environmental conditions, such as temperature and wind speed, so that they are optimal for photosynthesis, biosynthesis and growth;

- minimising losses of essential resources to other plants, particularly weeds;

- minimising losses of net primary production (NPP) and stored grain to pests like insects and rodents.

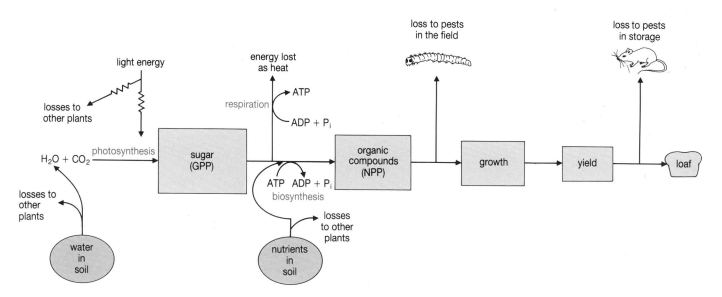

Fig 7.1 The challenge faced by farmers. At each stage of the production process, crops suffer potential losses in yield as a result of both abiotic (physical) and biotic (biological) factors.

The job of the agricultural ecologist is to devise agricultural practices and processes which achieve these objectives. We will examine how successful they have been in the next five chapters, in which we will examine the limits placed on agricultural production. In this chapter, we will consider mainly climatic factors, in the next two soil, and finally we will look at pest control.

7.1 THE PRINCIPLE OF LIMITING FACTORS

Think about a plant growing from a seed during the spring. Initially, the environment is cold and temperature limits growth. As the temperature rises the next factor likely to limit the growth of the seedling is the amount of light it can absorb. This depends on the size of the plant's leaves. When its leaves are very small the energy input into the plant is low. As leaf area increases the days have become longer and the light becomes more intense, so light input to the leaves no longer limits growth. As the plant's leaves have been developing, so has its root system; the plant is trying to increase its water and nutrient intake to keep pace with its increased photosynthetic capability. Thus the availability of water, or any one of the essential plant nutrients, for example nitrogen or phosphorous, may now be limiting plant growth. Finally, if the availability of all these nutrients is kept above some limiting level the final limits to growth may be imposed by the availability of light or, just possibly, by the low concentration of carbon dioxide in the atmosphere. There are three messages to be understood here.

1. Plant growth requires nutrients and energy in appropriate proportions.
2. The limit to growth at any particular instant is set by the factor which is in short supply – the **limiting factor**.
3. The limiting factor changes with time.

Understanding the principle of limiting factors will help you to explain many of the agricultural practices which we will examine in this theme. A farmer is often trying to remove a limitation acting on a crop which is actually growing, or trying to prevent a limitation acting on a crop which he intends to grow. The problem for us, as ecologists, is that these limiting factors, light, water, nutrient supply and so on, do not act in isolation but continually interact with each other in affecting plant growth. What that means, in terms of this book, is that while you will read about each factor in turn, as though they were acting in isolation, you must bear in mind that they continually interact with each other to produce their effect on plant growth.

Another important point to remember is that farming is a business and that farmers are subject to economic limitations. There is some optimum level of money which is worth spending on growing a crop. What this optimum is will depend on the value of the crop and the fixed costs of production. Thus, even if farmers could make their plants grow like Jack's beanstalk, they are unlikely to try because the costs involved would make the farm unprofitable. To emphasise this important aspect of farming consider the following two examples. They will give you more practice at understanding the principle of limiting factors, plus some useful additional information.

Soil temperature

The six months from April to September are a fairly well-defined growing season in Britain. Crops which ripen in the late summer could give heavier yields if the growing season started earlier. But the start of the growing season is determined by soil temperature. Too low a soil temperature will not allow germination of seed. One way to remove this limitation would be to heat the top few centimetres of the soil for a few weeks before April. However, the cost of doing this, given the high cost of electricity, would be enormous. Just as underfloor central heating of fields is a silly idea, so is flood lighting of daylight intensity to maintain growth at night, and burning coal all around the field to increase the carbon dioxide content of the air.

The climate, which determines soil temperature, length of daylight and rainfall is largely something which farmers just have to live with. The sensible thing to do is to choose crops and cultivation practices which will maximise production under the prevailing climatic and economic conditions.

Nutrients

Suppose a farmer plants wheat in a field whose soil contains too little phosphorous. Even if the wheat's requirement for water, nitrogen, potassium and other nutrients are met, the wheat will grow poorly when it has used up the available phosphorous. Here the availability of phosphorous is the limiting factor that determines how much wheat the field will yield. However, in this case the farmer can do something about the problem by applying a fertiliser. Such fertilisers, although not cheap, will increase the yield of a crop sufficiently to make their use economically viable.

QUESTION	7.1 (a) Using Fig 7.1 identify six further factors which may limit crop yield.
	(b) Classify your factors as either biotic or abiotic.
	(c) Which of the factors that you have listed does a farmer simply have to live with and which can be manipulated?

7.2 AGRICULTURE AND CLIMATE – AN INTRODUCTION

On a global scale, agricultural productivity depends on four major factors – light, temperature, water and nutrients. The amount of light falling on a crop is the ultimate limiting factor but, as we have seen and will continue to see, all four factors interact. Crops only grow well at temperatures above 5°C and if there is enough soil water to prevent wilting. Furthermore, plants can only grow at their maximum rate if their stomata are fully open, ensuring unrestricted gas exchange. A restricted water supply will cause the stomata to close, before wilting occurs, reducing the supply of carbon dioxide to the chloroplasts, so reducing photosynthesis and, ultimately, growth.

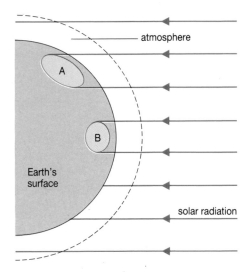

Fig 7.2 Solar radiation is spread over a greater surface area at higher latitudes (A) than in equatorial latitudes (B). This means that both light intensity and temperature will increase towards the equator.

Fig 7.3 Cloud cover over the Earth. What you notice most from this photograph is the absence of cloud cover over the Sahara desert and Mediterranean.

Growing season

The first choice a farmer has to make is which crop to grow. This will be largely deteremined by the length of the growing season – a period determined by the linked climatic factors of light availability, temperature and water supply. Temperature and light intensity vary in a systematic way with latitude as shown in Fig 7.2. Essentially, the further you are from the equator the less intense the incoming solar radiation (intensity is measured in kJ m^{-2}), so the lower the temperature and the light intensity.

Water availability

Fig 7.3 suggests that the distribution of cloud cover and hence precipitation (rain, snow, hail and so on) is irregularly distributed over the globe. However, the important point for a plant is the supply of water from the soil which is not related simply to precipitation. For example, if the soil temperature is low the rate of water uptake by the roots may be too low to keep pace with the rate of water loss by transpiration through the leaves, especially if the air temperature is high. Similarly, at high temperatures, or under windy conditions, the rate of transpiratory water loss may be so great that the plant simply cannot absorb enough water from the soil fast enough to make good the losses. In all cases the stomata close and photosynthesis slows down.

To overcome these problems, we need some index of water availability to plants which takes into account these interactions between climatic factors. **Potential evapotranspiration (PET)** is the theoretical amount of water which would evaporate from a dish of water in a field. It takes into account the effects of temperature, wind speed and relative humidity on evaporation. The relationship between PET and precipitation can then be used as an indicator of water surplus or deficit in the soil, as shown in Fig 7.4 for a desert.

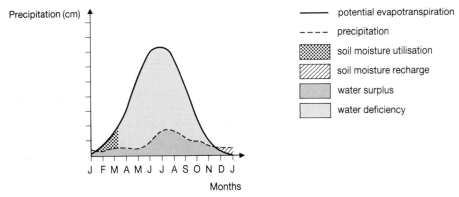

Fig 7.4 The relationship between precipitation and potential evapotranspiration in a desert. The vertical extent of the water deficiency (shown in pink) indicates the severity of water limitation.

Actual evapotranspiration (AET) is a measure of the combined loss of water from the soil by (1) evaporation of water from the soil surface and (2) transpiration from the leaf surfaces. It is calculated from data on precipitation and temperature and published as tables on a global basis. Do not confuse PET and AET. They will only be similar under conditions of high temperature and rainfall.

Agroclimatic zones

Using criteria of temperature, PET and AET the world can be mapped into the basic agroclimatic zones shown in Fig 7.5. Note that two of the zones,

CLIMATIC FACTORS AFFECTING PRODUCTIVITY

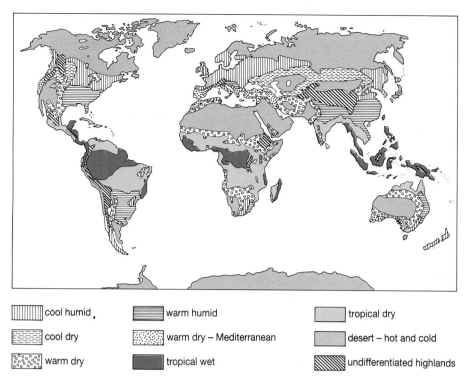

cool humid

cool dry

warm dry

warm humid

warm dry – Mediterranean

tropical wet

tropical dry

desert – hot and cold

undifferentiated highlands

Fig 7.5 The world can be mapped into agroclimatic zones on the basis of temperature and rainfall. Certain crops will only grow in certain climatic zones.

Fig 7.6 Growing crops in arid regions like this one is difficult, so life revolves around grazing animals like these cattle which can survive on the sparse natural vegetation. However, here, overgrazing has led to the destruction of the vegetation causing problems of soil erosion.

Table 7.1 Areas of the USA where plant growth is subject to limitations of various forms from the environment or from the soil.

Limitation	Area affected (%)
drought	25.3
shallow soil	19.6
cold	16.5
wet	15.7
alkaline soil	2.9
salinity	4.5
other	3.4
no limitation	12.1

desert and highlands, are too inhospitable for useful agricultural primary production although they may provide some limited grazing for livestock (Fig 7.6).

Given that raising temperatures over large areas is impracticable, Table 7.1 shows that the removable limits to production vary in different areas. For example, in Europe, which apart from the Mediterranean is a wet area, countering a lack of nutrients is the first priority. On the other hand, Table 7.1 shows that water shortage is the main problem in the United States. As we will see in the next chapter, civil and chemical engineering can overcome both these problems. For the rest of this chapter, we will take a closer look at three other factors – temperature, light and wind.

CLIMATIC FACTORS AFFECTING PRODUCTIVITY

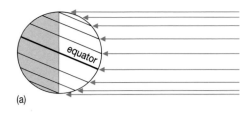

(a)

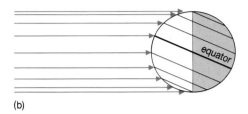

(b)

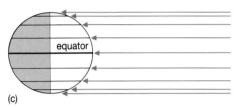

(c)

Fig 7.7 The position of the Earth on **(a)** 22 June: summer in the northern hemisphere bringing long, warm days and short nights; **(b)** 22 December: the northern winter with short, cold days and long nights; **(c)** 21 March and 23 September: spring and autumn begin in the northern hemisphere with days and nights of equal length. The seasons are reversed in the southern hemisphere.

Air temperature

The basic distribution pattern of annual mean temperature is related to latitude: because the Earth's surface is curved the amount of solar radiation which falls on a given area of ground per unit time is greater at the equator than at higher latitudes, e.g. in Britain (see Fig 7.2). However, this basic latitudinal distribution of temperature is modified by a number of other factors.

Seasons

Latitudinal and seasonal variations in temperature cannot really be separated. Look at Fig 7.7. This shows that the Earth is tilted at an angle relative to the Sun which affects the amount of sunlight striking different parts of the earth depending on the season. This effect produces the generalised temperature zones shown in Fig 7.5. You should realise that these zones reflect yearly average temperatures; the hottest temperatures actually occur in middle latitudes, not at the equator. For example, practically everywhere in the United States has experienced a summer temperature greater than 38°C – while Belem on the equator in Brazil has never experienced a temperature greater than 35°C.

Other factors also affect temperature distribution.

Aspect: South-facing slopes in the northern hemisphere receive more solar radiation than north-facing ones, hence they are warmer and may be more suitable for growing crops like grape vines.

Aspect: South-facing slopes in the northern hemisphere receive more solar radiation than north- facing ones, hence they are warmer and may be more suitable for growing crops like vines.

Continentality: Water heats up and cools down more slowly than land so that large bodies of water tend to moderate the temperature around the edges of the land masses surrounding them producing 'oceanic' climates. By contrast, the interiors of continents show much more marked seasonal changes of temperature, with the hot summers and very cold winters typical of a 'continental' climate. Thus London, which is further north than Prague, has milder winters because of the moderating influence of the sea.

Air and water movements: The coldest periods in Britain usually occur when there are north or north-east winds from the Arctic or eastern winds blowing from Siberia. Warm weather, on the other hand, is often accompanied by a southeasterly wind blowing from the Mediterranean. Just as winds can move cold air, so ocean currents can move large masses of warm and cold water, so modifying the climates of islands and continental edges which lie in their path. Thus the northward extension of the cool humid zone in Europe (see Fig 7.5) is due largely to warm water brought from the Gulf of Florida by the Gulf Stream and the North Atlantic Drift.

These modifying factors affect both the average yearly temperature and the annual mean range of temperature. Over most of the tropics this range is less than 7°C but it may be more than 75°C in parts of Siberia. Even more important for the farmer is local microclimatic variation. For example, the sinking of dense, cold air into the bottom of a valley at night can make it 3°C colder than the side of the valley only 100 m higher. Even slight indentations in the land can produce areas where the temperature drops low enough for ground frost to form even though the surrounding land is above freezing point. Crops planted too early in such frost hollows may be severely damaged. Air temperatures are a poor indicator of the temperature actually experienced by a plant since the presence of plants modifies the microclimate around them (Fig 7.8). So while data on variations in air

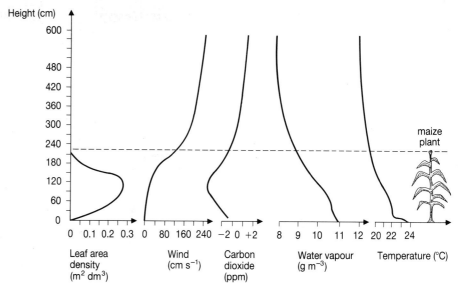

Fig 7.8 A crop modifies the microclimate around each plant. This diagram shows how certain climatic features change within a maize crop.

temperature from different sites provide a means of comparing different areas, they do not show the actual temperatures experienced by plants.

ANALYSIS

Worlds apart
This exercise requires you to use library research and thinking skills.

Britain and the Kamchatka peninsula in the eastern USSR lie at the same latitude and are of similar size. The mean winter temperature in Britain is 0°C to 8°C whilst in Kamchatka it is –16°C to –24°C inland and –8°C to –16°C on the coast. Mean summer temperatures are similar, as is the total annual precipitation although in Britain this is distributed throughout the year while in Kamchatka it occurs mainly in the summer. Using your library and the knowledge you have gained in this section account for these observations.

Soil temperature

The variations in microclimate noted above also apply to soil. Increasing soil depth has two effects on fluctuations in soil temperature.

1. The fluctuations are reduced or damped. A metre below the surface, daily air temperature fluctuations of many tens of degrees are damped out. At a depth of a few metres, even annual fluctuations are damped out. Thus parts of Siberia, which experience an annual variation of 75°C in air temperature, have soil which is permanently frozen, **permafrost**, just a few metres below the surface of the ground.
2. Soil temperature fluctuations lag behind those in air temperature. Thus the soil warms up more slowly in the spring but retains its warmth longer in the autumn than the air.

This temperature buffering capacity of soils is largely a consequence of their moisture content. Water has a high specific heat capacity which means that it can absorb a lot of heat energy before its temperature changes. Despite this buffering capacity, soil temperatures at seed-sowing depth (10–50 mm) can range between 3°C and 25°C during the course of a single day during early May in northern Britain. The implications of this for crop growth will become evident in the next section.

7.2 Fig 7.9 shows the annual course of air temperature and daylength at a site in Britain. Explain
  (a) the general shape of each curve,
  (b) the fact that the curve for air temperature lags behind that for light intensity.
  (c) How would you expect soil temperature to vary over the course of the year at this site? Explain your answer.

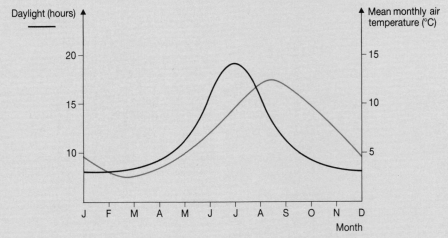

Fig 7.9 Variation in air temperature and daylength at a site in Britain.

7.3 Moselle is a major wine growing area in Germany. The river Moselle, which flows from west to east through the Hartz Mountains of Germany, has a steep-sided valley. Explain why the grape vines grow best on the north side of the river.

7.4 (a) What do you understand by the terms 'mean annual temperature' and 'mean annual temperature range'?
  (b) Explain the following observation: the mean annual temperature range increases with distance from the equator and the sea.
  (c) What are the implications for the statement in **(b)** for the length of the growing season in (i) Britain (ii) the middle of the United States?
  (d) Given your answer to **(c)** suggest why the mid-west of the United States is a major producer of maize while Britain is not.

7.3 TEMPERATURE AND PLANT GROWTH

Temperature directly affects crop growth through its effect on:
- metabolic processes, e.g. photosynthesis and respiration;
- seed germination, flowering and tuber formation;
- freezing and chilling injury.

The effects of temperature on plants is inextricably linked with problems of water supply. At high temperatures, water evaporates rapidly from the surface of leaves, which acts as a cooling mechanism but is highly undesirable if water is in short supply; at low temperatures, water may become unavailable because it is frozen.

Cardinal temperatures

Every plant has an upper and lower lethal temperature above or below which it suffers irreversible damage and dies. This fact alone restricts the choice of crops a farmer can grow: you cannot grow coffee in Britain because it is too cold. In addition, each crop has optimum temperatures for its development although this may vary for different parts of the same plant. These cardinal temperatures may vary depending on the stage of development of the plant or the time of year. For example, the woody branches of sycamore have a lower lethal temperature between –30°C and –50°C in the winter but –3°C to –4°C in the summer. The development of such **winter hardiness**, which is common to many trees, in the autumn and its subsequent loss in the spring is an example of the process called **acclimatisation**.

Photosynthesis and respiration

As temperature increases so does the rate of gross photosynthesis (GP) up to a certain point, after which it levels out (Fig 7.10(a)). By contrast, respiration (R) tends to increase almost exponentially with temperature (see Fig 7.10(a)). The result is that net photosynthesis (GP – R) is maximal at temperatures well below those for gross photosynthesis (see Fig 7.10(a)). Indeed, at temperatures above about 30°C respiration may well exceed gross photosynthesis, so reducing plant growth. At even higher temperatures the enzymes controlling metabolism will begin to be affected. These effects of temperature on metabolic rate are transposed into effects on growth (Fig 7.10(b)).

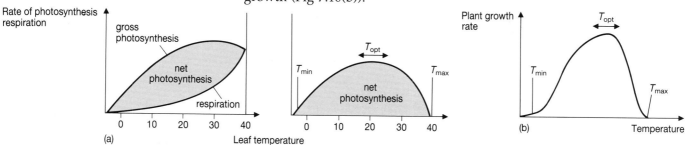

Fig 7.10 A diagrammatic representation of a plant's response to temperature. **(a)** The influence of temperature on gross photosynthesis, respiration and net photosynthesis in a typical plant. **(b)** A generalised diagram of the response of plant growth rate to temperature, illustrating three critical temperatures (T_{min} = minimum temperature, T_{max} = maximum temperature, T_{opt} = optimum temperature for growth).

Seed germination

For each plant species there is some range of temperatures over which that plant's seeds will germinate. For example, barley, wheat and rye seeds will not germinate below 3–5°C or above 40°C; maize grains will not germinate below 8°C or above 44°C. Here, then, is an immediate limitation on the length of the growing season. Barley seed, for example, planted in the spring will not germinate until soil temperatures are above 3–5°C. Some seeds, for example those of celery, need to be exposed to fluctuating temperatures before they will germinate. Others, like those of apples, cherries, peaches and plums, need to be exposed to low temperatures under moist conditions for several weeks or months. Such seeds are treated commercially by putting them in layers in moist sand and chilling them – a practice called **stratification** – to increase the percentage germination.

Vernalisation

Some plants need a period of exposure to low temperatures before they will flower. These include biennial plants like sugar beet which grow vegetatively in their first year and flower in their second year after exposure to winter cold. This means that seedlings of sugar beet and other biennial species grown for their roots, e.g. turnips, must not be exposed to low temperatures since this would cause the plant to flower (called **bolting**) in its first season of growth, before the root is harvested. This would reduce the quality of the root. The farmer must not therefore sow these crops too early since late frosts could vernalise them. Some annual crop plants, like winter varieties of wheat and rye, will flower if grown continually at high temperatures but do so sooner if exposed to temperatures between 3°C and 8°C. So the varieties of cereals known as winter annuals, like winter wheat, are sown in the autumn and begin to grow before the onset of winter. During the winter the young plants are exposed to low temperatures. They start to regrow in the early spring and flower in the early summer. The advantage of this to the farmer is that the plants are making the best use of the available light early in the growing season.

Freezing and chilling injury

Injury by frost is a major cause of damage to crops. The formation of ice crystals inside cells causes mechanical damage to cell membranes and walls but, more importantly, the ice crystals absorb cell water leaving behind a highly concentrated cell solution which may damage cell structure. Young seedlings in particular are susceptible to frost damage, which again places a major limitation on the growing season; you dare not plant some crops until all danger of frosts is past.

Young plants may acquire low temperature tolerance if they are exposed gradually to low, but not lethal, temperatures. This acclimatisation process is the basis of the horticultural practice of 'hardening off', used with seedlings raised early in the season under glass. For example, pea seedlings kept at 5°C for a few hours on several successive days can then withstand a temperature of −30°C without injury, a temperature which would kill untreated plants.

Prevalence of late spring and early autumn frosts may make some areas unsuitable for some crops. For example, peaches grow well in the Niagara peninsula near Lake Ontario in Canada. They do so because of the moderating influence of this large lake on air temperature. Thus by the time it is warm and the plants have acclimatised and lost their winter hardiness there is little chance of a late frost. By contrast, further south, away from the moderating influence of Lake Ontario, even though the climate on average, is warmer there are more late frosts and so no peaches. At the other end of the temperate growing season early autumn frosts may damage crops like maize which, in a country like Britain, require a long growing season to mature.

Temperatures can be lethal without dropping as low as freezing. If the temperature is low enough, metabolic reactions slow down and virtually stop. While this may not be serious in the short term, if the exposure to low temperatures continues for long enough it may weaken the plant to the extent that it may die. This phenomenom in plants is called **chilling injury** and often occurs at temperatures around 10°C. The probable cause is a disruption of membrane structure, although this often manifests itself as a malfunction in the plants' capacity to take up and retain water.

QUESTIONS

7.5 (a) Suggest evolutionary explanations for (i) acclimatisation; (ii) vernalisation; (iii) the requirement for seeds to be chilled for long periods before they will germinate. You will find it useful to define the terms and give some examples.

(b) When is vernalisation (i) useful (ii) a nuisance to the farmer?

(c) Why is it important to sow the seeds of winter wheat as early in the autumn as possible? What process is occurring in the young wheat plants which enables them to withstand winter cold?

7.6 In what sense can farmers be considered to manipulate the thermal environment experienced by their plants?

Sunlight consists of a continuous spectrum of wavelengths of light, only some of which, photosynthetically active radiation (PAR), can be captured by the photosynthetic machinery of the plant. So from the plant's point of view three aspects of light are important.

1. **Light intensity**: This is the amount of solar energy falling on a unit area of crop per unit time and is mainly affected by the angle of the Sun above the horizon, the **solar altitude** (Fig 7.11), and cloud cover.

2. **Light duration** (i.e. daylength): This determines the time that light is available for photosynthesis. Since daylength and maximum solar altitude are both affected by the time of year, the month with the longest days, June in Britain, is also the period with the highest light intensity. So the total amount of light available to a crop (light intensity × daylength) will be greater in the summer than in the winter in temperate regions.

3. **Light quality**: Many of the events in a plant's life, for example flowering, are determined by internal biological clocks many of which monitor daylength or photoperiod. Photoperiodism is discussed in another book in this series, *Biology*, by M. Rowland.

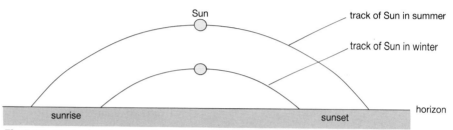

Fig 7.11   The track of the sun in the sky in Britain during the winter and summer months. Note how the daylength and the height of the Sun in the sky, the solar altitude, increases in the summer resulting in an increase in both light duration and light intensity.

Photosynthesis and canopy structure

The relationship between light intensity and the rate of photosynthesis is shown in Fig 7.12. You should already be familiar with this graph and the concept of compensation point. Now leaves at the top of the canopy of a crop during the middle of a bright summer's day will actually be receiving more light than they can actually use. However, lower down in the canopy

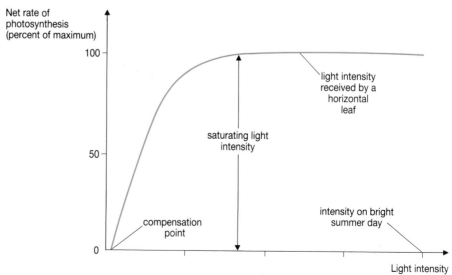

Fig 7.12   The relationship between light intensity and the net rate of photosynthesis (gross photosynthesis–respiration).

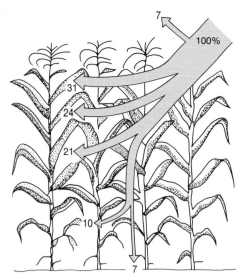

Fig 7.13 Light attenuation in a maize crop. Numerical values are percentages of the light intensity at the top of the canopy.

the light intensity falls dramatically (Fig 7.13). Leaves do not usually overlap completely so that flecks of full sunlight may still penetrate through small gaps in the upper canopy to reach the lower canopy. However, much of the light reaching the lower canopy will have passed through one or more leaves. The amount of sunlight absorbed by a leaf depends on its chlorophyll content but is usually about 90%. So the second layer of leaves will only receive 10% of the sunlight falling on the leaves at the top of the canopy while the third layer of leaves will only receive 10% of 10%, i.e. 1%, of the light falling on the top of the canopy. In other words it gets dark pretty quickly the further down in the canopy you are.

Imagine that you are a leaf at the bottom of the canopy. It is so dark here that you are likely to be near, or even below, your compensation point, i.e. you are respiring faster than you are photosynthesising. You are therefore contributing very little to the plant. Indeed, you may be a drain on the plant's resources so the plant would be better off without you. Such leaves usually senesce, turn yellow and fall off. Clearly, then, to achieve maximum yield a farmer needs to manage the crop so that it has the maximum amount of photosynthetically active leaves per unit land area. We will examine two aspects of this management problem.

Leaf orientation

Imagine the leaves at the top of a canopy are orientated obliquely to the incoming sunlight (Fig 7.14) rather than at right angles to it. This oblique angle reduces the intensity of the light per unit leaf area but this does not reduce the rate of photosynthesis in a leaf at the top of the canopy since light intensity is already above the optimal level (remember we are talking about the middle of a bright summer's day; refer to Fig 7.12). When leaves are slanted we can pack more of them into the top layer of the canopy. So more of the crop's leaf area will operate at optimal efficiency and growth will increase.

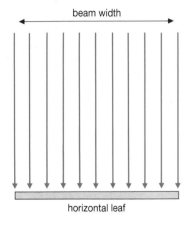
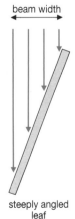

Fig 7.14 Angled leaves absorb less light than flat ones leaving more light to penetrate further into the plant's canopy. The upper leaves of plants that use sunlight efficiently are therefore usually slanted while the lower ones are flat to absorb the maximum amount of solar radiation.

wild beet

cultivated beet

Fig 7.15 Selective breeding of beetroot has produced the modern cultivated plant with its more upright leaves. This arrangement of leaves is more efficient in intercepting sunlight than its wild ancestors.

Now think about a leaf below this canopy of slanted leaves. Remember the light intensity is now only a fraction of that falling on the top of the canopy. The lower leaves will make the best use of the available light by being at right angles to it, so capturing the greatest amount of light per unit leaf area. The ideal plant for a farmer would therefore have its lower leaves horizontal, with the leaf angle increasing at each successive layer in the canopy, to almost vertical in the top layer. Such an arrangement of leaves can be achieved by selective breeding (Fig 7.15).

CLIMATIC FACTORS AFFECTING PRODUCTIVITY

Photosynthetic capacity

The growth rate of a crop depends on the size and efficiency of the photosynthetic system. The leaf area index (LAI) gives the size of the system while the net assimilation rate (NAR) gives its efficiency. You should remember these terms from the last chapter. The highest rate of production occurs when the product of LAI and NAR is at a maximum: crop growth rate = LAI × NAR. So any factor which alters either LAI or NAR will affect crop growth rate.

NAR depends primarily on light intensity and duration. Leaf area is determined by the size of the crop, which depends on:

- how long the crop has been growing – determined by the date of planting for annual crops like wheat (see Fig 6.11);
- how fast the crop has been growing – in the seasonal environment of Britain this will depend primarily on the temperature and leaf orientation.

Table 7.2 demonstrates the effect of sowing date on the yield of sugar from a British beet crop. Clearly, the earlier you can sow the better. However, we saw earlier that soil temperature imposes a limit to this. This example highlights a major limitation facing farmers in temperate areas like Britain. Ideally, you want the crop to have its maximum leaf area when light intensity and light duration are at their greatest, i.e. in midsummer, and when NAR is at some optimum value. But plants require time to grow. Cold spring weather limits how early a crop can be sown. Often the crop has not developed its maximum leaf area by midsummer when light availability is greatest, i.e. LAI is less than optimal for the amount of light available. So even though NAR is optimal the canopy is too small to intercept all the available sunlight which, in a monoculture, will fall on the soil and be wasted.

Now consider the same crop in July. The days are starting to shorten, light is less intense and so NAR starts to fall. However, many crops are still not ready to harvest. Some crops like sugar beet and kale continue to produce new leaves so that maximum LAI is not achieved until late in the growing season (see Fig 6.11), a time when external factors are becoming less favourable for photosynthesis. For example, when sugar beet reaches its maximum LAI in late August the light available for photosynthesis is only 70% of that in mid-June. The extra leafiness of the crop, by shading the lower leaves, may further reduce NAR; the mature plants are too leafy for the amount of light which is available. Thus in temperate areas like Britain the seasonal changes in light availability are poorly matched by changes in light interception by crops as they grow. This is primarily the result of the delay in the start of the growing season due to low spring temperatures, particularly those in the soil.

Table 7.2 Sowing date and yield of sugar from a beet crop in the UK.

Date of sowing	Sugar yield (t ha^{-1})
28 March	10.5
8 April	10.6
22 April	9.5
5 May	8.8
19 May	7.0

Fig 7.16 A young crop of sugar cane intercropped with peanuts.

Light management

There are three ways in which to tackle this problem.

1. Sow the crop in the previous autumn, before the soil cools down too much, so that the plants germinate before winter. The crop then overwinters as a small plant which is ready to start growing again as soon as conditions improve in the spring. This is the idea behind growing winter cereals.
2. Increase the density of sowing. A more dense crop of seedlings will intercept more light than a thin crop. However, with increasing density, intra-specific competition between the crop plants will tend to produce an approximately constant harvested crop irrespective of sowing density. It makes no economic sense to sow expensive seed too densely.
3. Use a system of intercropping (Fig 7.16). This method, practised on some 40% of the world's cropped areas, relies on the fact that some crops, like

radishes, will grow more rapidly than the crop they are planted between, for example lettuces. Thus the radishes will be ready for harvesting when the lettuce plants are just reaching their maximum LAI and would start shadowing the radishes. The two crops together make more efficient use of the available sunlight than lettuces would on their own. Such intercropping may also be effective in preventing some pest problems. For example, planting onions among carrots may mask the 'carroty' smell which attracts carrot root fly.

QUESTIONS

7.7 (a) Fig 7.17 shows some data for light intensity for three different sites. Which site is furthest north, nearest the equator and in temperate latitudes?
(b) Explain why light availability in Britain will be greatest in the summer and least in the winter.

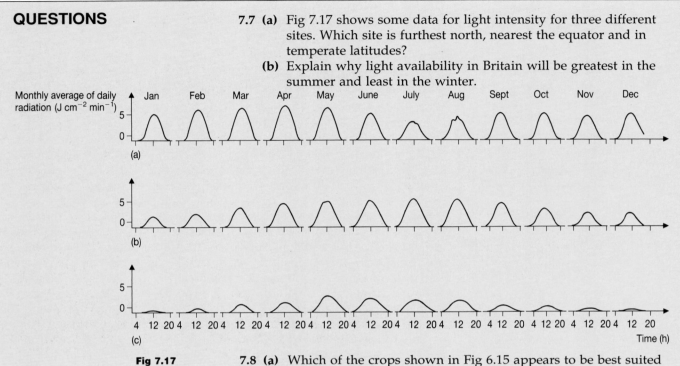

Fig 7.17

7.8 (a) Which of the crops shown in Fig 6.15 appears to be best suited to the conditions in temperate regions?
(b) Would a wheat or a sugar beet crop have a greater optimal LAI? Explain your answer.
(c) In terms of leaf area development what would constitute an ideal crop?
(d) How might breeding cold-tolerant varieties of crops, with seeds which germinate at low temperatures, increase yields.
(e) Explain why the yield of grain from cereals with upright leaves, like wheat, is higher than that of maize which has leaves at right angles to its stem. Remember that yield is measured in units of mass (e.g. tonnes) per unit area.

7.9 Recently, the use of transplanted seedlings of sugar beet, raised under glass prior to planting in the field in March, has been shown to produce 9% more sugar per hectare than from a comparable crop grown from seed sown in the field in March. Sugar beet is a biennial which only stores sugar in its root prior to flowering.
(a) Why should sugar beet not be sown in the field before the end of March?
(b) Explain why a higher yield is obtained from transplanted seedlings.
(c) Why might farmers continue to sow seeds despite the increased yield obtained from transplanted seedlings?

7.10 Suggest reasons why intercropping is not widely practised on a commercial scale in Britain or the United States.

7.5 WIND

Crops are affected by wind at both a structural and physiological level. Strong winds and rain may flatten cereal crops, a process called **lodging**, which makes them difficult to harvest, or may even cause the grain in contact with the ground to rot. Taller plants buffeted by the wind may suffer damage to their waterproof cuticles, so increasing their rate of water loss. Wind may also blow soil away creating massive problems of soil erosion (Fig 7.18).

Fig 7.18 The beginning of a dust storm in the Arkansas River region of the great flat Oklahoma Plain in the mid-west of the United States.

Physiologically, wind affects the temperature and water relationships of plants. All plants, and animals, are surrounded by an insulating boundary layer of comparatively still air. The rate at which heat, gases and water vapour are transferred across this boundary layer is inversely proportional to its thickness. Blow on the back of your hand. You should notice the cooling effect. This is happening because you are reducing the thickness of the boundary layer on the surface of your skin. This increases the rate at which heat is transferred across the boundary layer from your warm body to the cooler air. The effect will be the same for a plant in a wind, provided the air temperature is greater than that of the plant.

However, the temperature of a body is also affected by the rate of evaporation of water from it. As the rate of evaporation (i.e. transfer of water vapour across the boundary layer) is influenced by boundary layer thickness, and therefore by wind speed, there is inevitably an interaction between the effects of wind on temperature and on water loss. We really cannot get away from these interactions. They make life difficult for us but they are a reality for the crops the farmers are trying to grow!

The effects of such interactions are complex but the most usual effect of high winds is to cool and to increase the rate of water loss from the plant. To avoid desiccation the plant will close its stomata, so reducing carbon dioxide supply and the rate of photosynthesis. As a result, growth will be slowed. Providing wind breaks, for example hedges, and shelter belts of trees will help to minimise the effects of wind on crop growth. For example, if the shelter belt has a height H then there will be a significant reduction in wind speed for a distance of up to $40H$ from it while there will be virtual calm within $5H$ of the wind break. However, shelter belts will increase the shading of the crop, and change the relative humidity and temperature of the air near to it. Increasing the relative humidity can make some plants more susceptible to attack by pests and fungal diseases. Nonetheless, such shelter belts can increase productivity by up to 50% in addition to preventing soil erosion.

ANALYSIS

The importance of hedges
This exercise involves interpreting graphical information.

Measurement of changes in some climatic factors caused by the presence of a hedge are given in Fig 7.19. The values for each factor are as compared with the values obtained in an open field.

(a) (i) Which two environmental factors are affected the most by the presence of the hedge? (ii) At what distance from a 2 m high hedge would these factors be most marked? (iii) Why is the relative humidity increased near the hedge although evaporation from the soil is reduced?

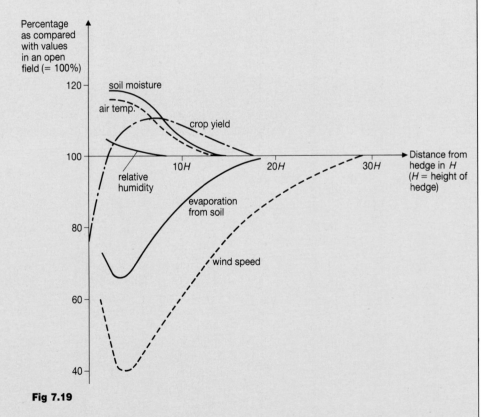

Fig 7.19

(b) Fig 7.19 also shows the changes in yield of crop plants near to the hedge. Account for (i) the decreased yield close to the hedge (ii) the increased yield between $2H$ and $18H$ from the hedge.

(c) (i) From the information given what would be the likely effect on the total crop yield of removing the hedge? State one piece of evidence to support your answer. (ii) Suggest two reasons why farmers may wish to remove hedges.

CLIMATIC FACTORS AFFECTING PRODUCTIVITY

7.11 (a) Define the term relative humidity.
(b) Why does washing dry better on a cold, windy day than on a hot, still, humid one?
(c) Will water loss from leaves be increased or reduced by increasing (i) the relative humidity (ii) the temperature of the air? Explain your answer. What effect will this have on the rate of photosynthesis?
(d) It has been noted that at low wind speeds the rate of net photosynthesis may increase. Try to explain this observation.

7.12 Consider the following two statements, the first of which is true. Is the second statement true or false and does it provide an adequate explanation of the first statement?
1. In some situations, crop yields are increased by an average of 10–30% immediately downwind of a shelter belt.
2. By reducing wind speed the shelter belt increases the boundary layer thickness and reduces the evapotranspiration of the crop.

SUMMARY ASSIGNMENT

1. Explain how the following factors might affect the productivity of a crop. Try and give your answer in terms of LAI, NAR, crop growth rate, time of planting and so on. (a) Time of the day and year; (b) the position of the leaves on the stem: for example bunched at the top or spaced down the stem; (c) the angle of leaves to the stem; (d) early spring temperatures; (e) the rate of transpiration; (f) the age of the leaves; (g) the amount of water in the soil in early spring.

2. Describe how the efficiency of interception of sunlight by the leaf canopy of a developing crop determines the growth rate and yield. In your answer define technical terms you use which relate directly to crop productivity and where possible refer to named crops.

3. Prepare essay plans to answer the following questions.
   (a) Discuss the meaning and relevance to crop production of the terms 'leaf area index' and 'leaf area duration'.
   (b) Using named examples, show how the development of the leaf canopy can be manipulated by the farmer to achieve higher crop yields.

4. In East Anglia, the main cereal growing area of Britain, large numbers of hedges have been removed to produce large fields. Investigate and report on **(a)** the reasons for this change, **(b)** any deleterious effects the change has had on (i) agricultural production (ii) the amenity value of the countryside.

Chapter 8

SOIL AND EDAPHIC FACTORS

Just think about soil for a moment. Even better go and pick up a handful. Squeeze it. Feel it between your finger and thumb. Examine it with a lens. You are holding (or thinking about) a truly remarkable material, a complex mixture of:

- inorganic materials – mainly clay, silt and sand;
- decaying organic material;
- water;
- air;
- living organisms.

The thin layer of soil, at the most 2m thick, provides the nutrients and water for the plants which, indirectly or directly, provide the food you eat, many of the clothes you wear, even the paper you write on. Yet, too often, we take all this for granted. Soil represents one of our most important, and most abused, resources. There is now considerable evidence that entire civilisations, for example those of ancient Mesopotamia, may have collapsed because they mismanaged the top soil essential to grow the crops needed to support their populations.

LEARNING OBJECTIVES

After completing the work in this chapter you will be able to:

1. describe the major edaphic factors;

2. describe the physical structure of soil;

3. appreciate that the physical structure of a soil results from the interactions between mineral fragments and organic material;

4. explain how a soil's porosity affects its water holding capacity and aeration;

5. classify soils.

8.1 THE NATURE OF EDAPHIC FACTORS

Ecologists call environmental factors operating in the soil **edaphic factors**. These interact in complex ways to determine the properties of soils. Before we look at each factor in detail it will be useful to have a brief summary so that we know where we are going. The five most important edaphic factors in agricultural soils are physical structure, water and air content, chemical compostion and soil pH.

Physical structure

This involves particle size, texture and overall structure of the soil. It is largely determined by the relative proportions of the different sized

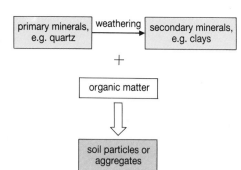

Fig 8.1 The mineral and organic components of the soil interact to produce soil aggregates.

Table 8.1 Mineral component of soil.

Size	Diameter (mm)	Visible using
gravel	>2.0	naked eye
coarse sand	2.0–0.2	naked eye
fine sand	0.2–0.02	naked eye
silt	0.02–0.002	light microscope
clay	<0.002	electron microscope

mineral particles (Table 8.1) and the way in which these interact with the organic matter in the soil (Fig 8.1). Physical structure affects the ease with which plant roots penetrate the soil, soil aeration, drainage and the capacity of the soil to absorb and hold water and mineral nutrients. Water in the soil is not in a pure state; it contains dissolved organic and inorganic compounds like nitrates and phosphates. Plants absorb the nutrients they need, usually by active transport, from this **soil solution**.

Water and air content

These two factors are inextricably linked. They depend primarily on the soil's porosity and the number and size of spaces it contains, which in turn is determined by the soil's physical structure. Pore spaces which are not filled with water are filled with air. Waterlogged soils, where all the pore spaces are filled with water, become anoxic (lack oxygen), so that anaerobic microbial processes, which produce toxic compounds like hydrogen sulphide, predominate, and so chemical conditions in such soils are very different from those in well aerated soils (Fig 8.2). Both the lack of oxygen in the soil and the changed chemical conditions reduce plant growth.

Fig 8.2 Some plants like rice, seen here growing in paddy fields in Malaysia, are adapted to living in waterlogged soils. The shallow water, which lies above the soil in which the rice plants are rooted, contains many nitrogen-fixing cyanobacteria ('blue green algae') which provide an additional source of nitrogen for the rice plants.

Chemical composition

This includes both nutrient concentrations and the chemical nature of the soil particles, which affects adsorption and availability of ions. In particular, clay particles and humus (see later) carry negative charges so they attract and hold ions with a positive charge. This prevents them from being washed out of the soil (**leached**) by rain and makes them available to plants. Table 8.2 lists the essential macro- and micro-nutrients for higher plants. An important point which must be emphasised is that the levels of mineral nutrients required for optimum growth vary between different plant species, as do the minimum and maximum levels tolerated (Fig 8.3). In addition, plants vary in their tolerance to high levels of potentially toxic non-essential elements like lead, aluminium and sodium. So different crops have different needs and tolerances for mineral elements in the soil. What will do for cereals, like wheat, will not suit brassicas, like cabbages and brussel sprouts.

Colloids and cation exchange in the soil

Solid particles less than 0.001 mm in diameter have properties intermediate between those of a suspension and those of a true molecular solution. This behaviour is related to both their large surface area to volume ratio and the electrical charges they carry. Such particles which behave in a colloidal way are called **colloids**. Important soil colloids include clay and humus.

Ions which carry a positive charge are known as cations. In the soil the millions of negatively charged colloidal particles of clay and humus attract the cations in the soil to balance their negative charge. Cation exchange between humus and clay minerals on one hand and the soil solution on the other is one of the most important factors in determining soil fertility.

Fig 8.3 This carrot has been grown in soils which are too rich in mineral nutrients and the result is a deformed growth. Potatoes grown in the same soil would, however, have flourished.

Table 8.2   Essential plant nutrients.
      * chlorosis is yellowing of leaves

Nutrient	Role in plant	Deficiency symptoms
Macronutrients		
Calcium	Component of middle lamella of cell walls; ties up waste products as insoluble salts; involved in membrane function	Meristem death; abnormal cell division; deformed tissues; breakdown of membrane structure
Magnesium	Component of chlorophyll and a cofactor of many metabolic enzymes	Chlorosis*
Nitrogen	Component of proteins, nucleic acids, chlorophyll and some hormones	Chlorosis of older leaves. Also severe stunting of growth
Phosphorus	Component of nucleic acids, ATP and phospholipids	Dark colour; loss of older leaves; stunting; slow development
Potassium	Important in maintaining membrane potentials and opening of stomata; activates enzymes in photosynthesis and starch synthesis	Mottled chlorosis, starting in older leaves; weak stems
Sulphur	Component of amino acids and coenzyme A	Chlorosis, poor root growth
Micronutrients		
Boron	Nucleic acid synthesis; pollen germination; carbohydrate transport	Thick dark leaves; malformations; cell division and elongation and also flowering inhibited
Chlorine	Used in production of O_2 during photosynthesis	Small leaves, slow growth and thick stunted roots
Copper	Component of some enzymes essential in respiration and photosynthesis	Dark misshapen leaves with spots of dead tissue
Iron	Chlorophyll production; part of electron transport molecules and of some enzymes	Chlorosis, appearing first in youngest leaves
Manganese	Activates citric acid cycle enzymes; involved in production of O_2 during photosynthesis	Mottled chlorosis
Molybdenum	Part of enzymes involved in nitrate reduction and nitrogen fixation	(as nitrogen deficiency)
Zinc	Synthesis of tryptophan, a component of some enzymes	Small puckered leaves; stem elongation reduced

Fig 8.4 illustrates the basic principles for potassium, an essential plant nutrient. Notice that Ca^{2+} and Mg^{2+}, two other essential plant nutrients, are also attached to exchange sites on clay particles. They are also subject to the same process of cation exchange with the soil solution.

Soil pH

Plants may be directly affected by pH but indirect effects are usually more important. In particular, pH influences the activity of soil microorganisms and the degree of dissociation and solubility of ionic substances. In this way it affects the availability of mineral ions (Fig 8.5). Again, the explanation lies in the cation exchange properties of soils.

Like potassium, hydrogen ions (H^+) can exist in solution, as well as in exchange sites. In normal soil an equilibrium exists between Ca^{2+} and H^+ in solution and Ca^{2+} and H^+ in exchange sites. As in the case of potassium, plants are continually removing the nutrient Ca^{2+} ions from the solution. This disturbance of the equilibrium is compensated by a release of Ca^{2+} from the exchange sites. This enables H^+ ions to leave solution and fill the exchange sites vacated by the Ca^{2+}. The effect of this is to increase the acidity of the soil. Such an acid soil contains fewer adsorbed cations in exchange sites for release into solution as nutrients.

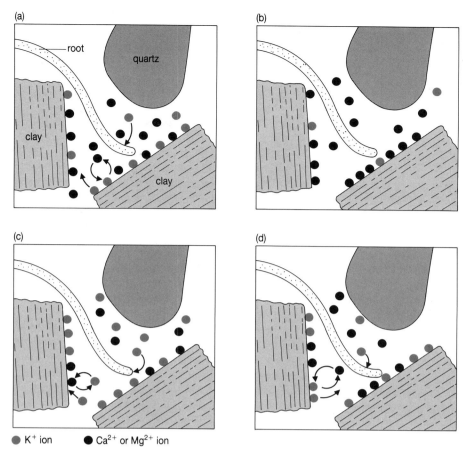

Fig 8.4 Cation exchange in the soil.

(a) The root takes up potassium from the soil solution, so potassium ions leave the exchange sites on the clay minerals to maintain equilibrium.

(b) Potassium deficiency results once potassium in the exchange sites has been used up.

(c) Excess potassium in the soil solution, due to the addition of potassium nitrate fertiliser, results in potassium being taken up into the exchange sites to maintain equilibrium.

(d) Steady release of potassium from exchange sites maintains potassium levels in the soil solution.

Key: ● K^+ ion   ● Ca^{2+} or Mg^{2+} ion

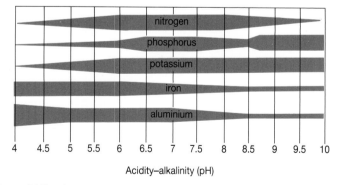

Fig 8.5 The availability of a nutrient is directly related to its solubility, which in turn depends on soil pH. The thickness of the horizontal red bands in the graph represents the solubility of each nutrient at the various pH levels.

Once again, then, we have a whole range of factors, covering a whole complex of soil conditions, which may interact with each other or vary independently. However, unlike climatic conditions, where the farmer could do little more than tinker with the environment (changing sowing dates, choosing appropriate crops, planting shelter belts and so on), agricultural practice can have a major impact on edaphic factors. Indeed, most

SOIL AND EDAPHIC FACTORS

of the things that farmers do (cultivating the soil by ploughing, discing and harrowing, liming the soil, applying manure and inorganic fertilisers, irrigating crops or laying drains) are directly aimed at altering edaphic factors to the benefit of the crop they are growing. Since many of these agricultural practices affect more than one edaphic factor at a time, we need to look at edaphic factors in a little more detail to understand the principles underlying the practice. Although we will look at one factor at a time, you should always bear in mind that they are continually interacting with each other to produce their effect on the soil and, hence, on plant growth.

QUESTIONS

8.1 Using the information in Fig 8.5 answer the following.
   (a) Which soils are likely to contain high concentrations of Na^+?
   (b) Why is the ideal soil pH for plant growth about 6.5?
   (c) Plants like rhododendron and azalea require large amounts of iron to grow well. Explain why these plants only thrive on acid soils.
   (d) On which soils is aluminium toxicity likely to occur?
   (e) Bearing in mind the charges on clay and humus particles can you explain why aluminium solubility increases as the pH decreases? (**Hint**: what makes a solution acid?)

8.2 Bulb fibre, used to grow bulbs in pots which do not have drainage holes, contains charcoal which absorbs hydrogen sulphide. Why is this a necessary precaution?

8.3 How could you work out the different nutrient requirements of different plant species?

8.2 THE PHYSICAL STRUCTURE OF SOILS

Soil, like a sponge, consists of a solid matrix and holes or pores. The nature of the matrix will determine the number, size and types of pores, which, in turn, will determine the soil's capacity to hold water and its degree of aeration. For example, a soil made of large particles will contain large pores through which water will drain easily. By contrast, a soil made of tiny particles will contain lots of small pores which will tend to retain water by capillary action.

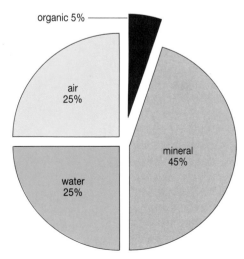

Fig 8.6 The percentage composition by volume of water, air, organic matter and inorganic (mineral) matter in an average soil.

Fig 8.6 shows the approximate composition of a good agricultural soil. Note that this soil consists of about half solid material (45% inorganic and 5% organic) and half space (pore space). At an optimum moisture level for plant growth about half this pore space will be occupied by water.

ANALYSIS

The answer lies in the soil
This exercise involves the use of observational skills.

The best way to study soils is actually to dig some up and look at it. Using a trowel, collect a sample of soil, ideally an intact vertical section from uncultivated ground. Examine the soil with a hand lens. In what form does the mineral matter, organic matter, water and air exist in your soil? Now take a small piece of the soil from the bottom of the section and roll it between your thumb and first finger. Use Table 8.3 to classify your soil. The meaning of the various terms will become apparent in a moment.

Table 8.3   Estimation of soil texture by 'finger assessment'.
* Plasticity refers to the ease with which the soil can be moulded.

Appearance under lens	Feel between fingers Damp or dry	Wet	Rolling between fingers	Texture
large grains absent or rare	smooth and non-gritty;	generally very sticky – plastic*	gives long threads which will bend into rings	clay; silty clay; sandy clay
many sand grains present	slightly gritty	moderately plastic	gives threads with difficulty which will not easily bend into rings	silty clay loam; clay loam; sandy clay loam
sand grains present but silt predominating	smooth	smooth	forms threads with broken appearance	silt; silt loam
comparable portions of sand, silt and clay	gritty	slightly plastic	gives threads with great difficulty	loam
sand grains predominate	more gritty	not plastic; only slight cohesion	gives threads with very great difficulty	sandy loam
mostly sand	very gritty	forms flowing mass	does not give threads	loamy sand; sand

Soil texture and structure

As a result of your investigation you should have put together a picture of soil like that shown in Fig 8.7. The mineral and organic components of the soil clump together to form soil particles or **aggregates** with air and water occurring in spaces or pores between the solid particles. The relative proportions of clay, sand and silt (see Table 8.1) determine the **soil texture** (Fig 8.8). However, these mineral particles do not exist as separate entities in the soil. Rather they are glued together into more or less stable aggregates with the organic matter. The distribution, shape and stability of the aggregates combine to make up the **soil structure** – one of the main factors determining porosity in soils. In particular, farmers are keen to encourage the development of the **crumb** or **granular structure** characteristic of good agricultural loams.

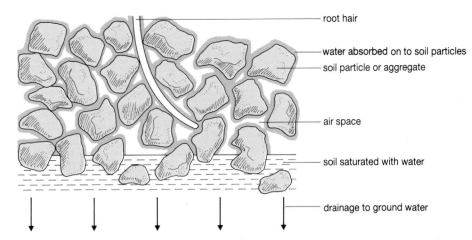

Fig 8.7 Fine structure of the soil showing soil aggregates, soil water and soil air. Each aggregate is about 100 μm in diameter.

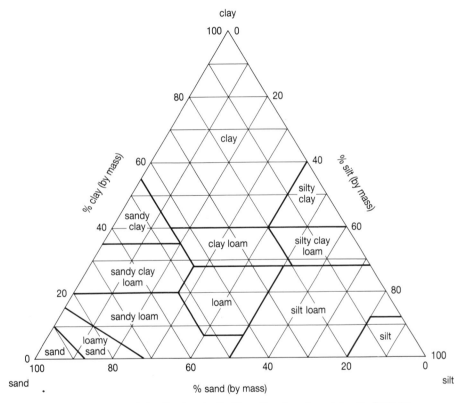

Fig 8.8 The textural classification of soils. Clay, silt and sand are defined in Table 8.1.

Organic matter

Although it may only constitute a few percent of the total soil content (see Fig 8.6) organic matter exerts an influence on the property of soils and consequently on plant growth which is out of proportion to its contribution to soil volume. In particular, organic matter:

- acts as a 'granulator' of mineral particles producing the loose, easily managed loam soils so prized by farmers (see Section 8.4);

- represents a source of plant nutrients, particularly phosphorus, sulphur and nitrogen;

- increases the amount of water a soil can hold and the proportion of that water which is available to plants;

- promotes good drainage and aeration of soils;
- serves as an energy source sustaining the soil microorganisms on which the soil's biochemical activity depends.

We can recognise two main types of organic matter in soil.

1. The original and partly decomposed remains of plants, animals and the products of their metabolism. This would include plant roots and leaves, the bodies and faeces of animals, and products of microbial synthesis. Such materials are used by soil organisms as sources of energy and materials for building new tissue. The activity of the soil organisms will also release nutrients from the organic matter which are then available for use by plants.
2. **Humus** which is a complex and rather resistant mixture of brown or dark brown amorphous and colloidal substances modified from the original tissues of plants or synthesised by soil microorganisms.

The importance of humus in the soil cannot be overemphasised. In particular, we need to examine the interaction between humus and clay minerals which is the major factor controlling the physical and chemical properties of soils. Since both clay and humus consist of extremely fine particles they have an enormous surface area in relation to their volume. These surfaces carry electrical charges so clay and humus particles form centres of activity in the soil around which chemical reactions take place. Since they attract ions to their surface they form reservoirs of essential plant nutrients, protecting them from loss by leaching and releasing them slowly for use by plants. The electrical charges also attract water molecules which can be bound so tightly to the soil colloids by hydrogen bonding that they cannot be taken up by plants. Finally, the surface charges allow clay and humus to form bridges between larger soil particles. This promotes the formation and stabilisation of a stable, granular soil structure.

Both soil temperature and soil aeration influence the rate at which organic material decomposes. Organic matter tends to accumulate in soils which are poorly aerated since this reduces the rate of aerobic respiration. The organic content of soils in cooler areas increases because of the reduced activity of soil microorganisms. Soils in warmer areas, for example the tropics, tend therefore to have less organic matter in them than soils formed in colder climates.

QUESTIONS	8.4 Distinguish between soil texture and soil structure.

8.5 Organic matter gives soil a brown or black colour.
   (a) What relationships might exist between soil colour and climate? Why?
   (b) Suggest why soils in tropical rain forests are relatively infertile.
   (c) Under what conditions would you expect peaty soils to develop?
   (d) Advertisements for some sorts of hover mower used to recommend that grass clippings be left on the lawn. Suggest why.

8.3 SOIL POROSITY

Fig 8.6 shows that 50% of the volume of a good agricultural soil is actually space occupied by water and air. The soil porosity – the number and size of pores it has – is a major factor controlling the amount of water and air the soil can hold and the rate at which water moves downwards (drains) through it. Soil porosity is in turn determined by the interaction between the mineral and organic components of the soil. Although we will consider them separately it is important that you realise the intimate relationship between soil water and soil air; you cannot change one without changing the other.

The pore spaces between the soil aggregates can be occupied by either water or air. In a particular soil the proportions of water and air are connected; as one increases the other decreases. For example, if the pores are filled with water, i.e. the soil is waterlogged, then the amount of air is greatly reduced. We will deal first with the soil solution. In particular, we need to know:

- how water moves in the soil;
- how much water the soil can store, its water-holding capacity;
- how much of this water is available for use by the plants.

Soil water

Water can exist in the soil in four states (Fig 8.9).

1. **Hygroscopic water** exists as a thin film held tightly around the soil particles.

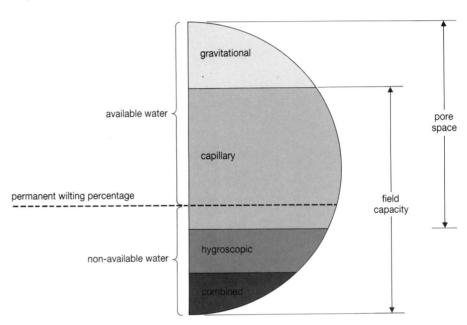

Fig 8.9 Forms of soil water. Note that a proportion of capillary water is unavailable to plants.

2. **Combined water** is chemically bound to the soil particles. (You can observe combined water by heating some hydrated copper(II) sulphate in a test-tube. What happens to the blue crystals as they are heated?) Neither combined nor hygroscopic water make the soil 'wet' and they can only be driven off by intense heat. Consequently, they are not available to plants.

3. **Gravitational water** is the water which occurs in the pore spaces which is potentially available to plants. However, very small pores hold water so tightly by capillary action that the water in them is unavailable to the plants. By contrast, very large pores, **macropores**, allow the water to drain away freely so the water in these pores, **gravitational water**, is only available to the plants for a short period of time after heavy rain or irrigation.

4. In between these two extremes are small pores, **micropores**, that can hold water against the action of gravity by capillary action (Fig 8.10) but are sufficiently large to allow plants to take up the water. It is this

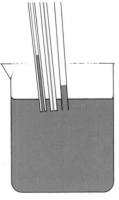

Fig 8.10 A demonstration of capillarity. The smaller the diameter of the tube, the further water rises up it.

capillary water, held in micropores, which supplies the vast bulk of the water that plants need.

Field capacity and wilting point

If you were to examine soil after heavy rain or after it had just been irrigated you would find all the pore spaces, large and small, filled with water. The soil, in this condition, is said to be **saturated** and any air in the pore spaces will have been displaced by water as it moves downwards into the soil. As the rain or irrigation continues water will continue to move further and further into the soil taking dissolved minerals, for example nitrates, with it. If enough water is applied it will eventually move into the underground **water table** (Fig 8.11). After it stops raining, water will slowly drain out of the macropores which will then fill with air. In a good agricultural soil this will take about three days and the soil is now at its **field capacity** (see Fig 8.9), which is given by:

Field capacity = capillary water + hygroscopic water + combined water + water vapour

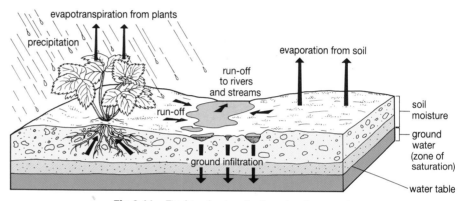

Fig 8.11   The fate of water after it reaches the ground.

The plants remove capillary water from the soil, losing it from their leaves by evapotranspiration. Capillary water will also evaporate directly from the soil. This lost water can be replaced by water moving up through the soil, from the water table, by capillary action. Even so the plants find it increasingly difficult to get water from the soil. Eventually, the supply becomes so low that the plants begin to wilt. If water is not added to the soil the plants will reach a point at which they cannot recover from wilting. The moisture content of the soil at this point is called the **permanent wilting percentage** (see Fig 8.9).

Soil air

The air in soil is also found in the pore spaces and differs from atmospheric air in three ways.

1. It is not continuous since it is separated by soil particles.
2. It usually has a higher moisture content.
3. It usually has a higher carbon dioxide and lower oxygen content (Table 8.4)

Table 8.4   Soil air composition.

Gas	Percentage Atmosphere	Soil
oxygen	20.99	20.3
carbon dioxide	0.04	0.5
nitrogen and other gases	78.97	79.2

The amount of air in the soil depends primarily on the amount of water in the soil. Air moves into soil pores not occupied by water. After heavy rain, air will move into the larger pores first as the water drains. As the soil dries out the air occupies pores of smaller size. So soils with a large percentage of large pores, sandy soils, which drain freely will be well aerated. By contrast, soils with a large percentage of small pores, such as clay soils, tend to drain poorly and therefore are poorly aerated. Since oxygen supply to the roots is essential for plant growth such soils may need structural improvement if they are to be used for agriculture.

QUESTIONS	
8.6	**(a)** What is the physical basis of capillarity?
	(b) Why does organic matter increase the water-holding capacity of soil?
8.7	**(a)** Why is the field capacity the sum of the capillary water, hygroscopic water, combined water and water vapour?
	(b) Using Fig 8.9 distinguish between available and unavailable water.
	(c) Is a soil which has reached its permanent wilting percentage completely dry? Explain your answer.
	(d) A proportion of unavailable water is capillary water. Why is this water unavailable for use by plants?
8.8	**(a)** Explain how biological activity affects the oxygen and carbon dioxide content of soil.
	(b) Why is aeration of soil important to plant growth?
	(c) Why might excessive watering harm plants?

8.4 MAJOR SOIL TYPES

We are now in a position to put all these different elements of soil structure together by looking at complete soils. What we are looking at is the soil in its uncultivated state. This will indicate what the farmer needs to do to the soil to improve its fertility. Soils dominated by large mineral fragments are gravelly or sandy. Where fine minerals predominate the soil is clayey. All gradations between these extremes are found in nature (see Fig 8.8).

Mineral and organic soils

The soil in your back garden or in the local park will consist mainly of mineral fragments, only containing between 1 and 10% of organic matter. However, soils in swamps, bogs and marshes often contain 80–95% organic matter. When drained such organic soils, for example the Fenland soils of East Anglia and the 'mosses' of West Lancashire, can be very productive. Alternatively, organic material, like peat, may be excavated, bagged and sold for use in the garden. However, despite their local importance, organic soils occupy only a tiny fraction of the land area compared with mineral soils.

Clay soils

Here the clay fraction of the soil exceeds 35%. If you think about modelling clay or walking across a clay field you will soon appreciate the main properties of clay soils. The small clay particles are easily packed together so when wet a clay soil is both sticky and plastic (easily moulded). When dry it becomes hard and cloddy. However, the high clay content of such soils means that they have large reservoirs of plant nutrients and therefore are potentially very fertile, if only their structure can be improved. Clay

Fig 8.12 This clay soil has been ploughed when the weather was too dry. As a result the soil has dried into these hard clods which will be practically impossible to break down by secondary tillage operations like harrowing.

soils have a low porosity and the pores are predominantly very small micropores. This means that:

- clay soils are poorly aerated;
- when dry, water cannot penetrate the soil easily;
- when wet, the soil drains poorly and therefore is prone to waterlogging;
- the soils warm up slowly in the spring since they contain so much water;
- they are difficult to work, i.e. clay soils are '**heavy**'.

Structural improvement of these fine-grained soils is not a simple problem. If heavy machinery is used on clay soil when it is wet the pore space becomes even further reduced and it becomes practically impervious to air and water. Any aggregates present are broken down and the soil is said to be **puddled**. When a soil in this condition dries it becomes hard an dense.

The tillage of such soils, ploughing, discing and harrowing, to produce the fine tilth needed to make a seed bed, has to be carefully timed (see Section 9.1). If you plough when the soil is wet, structural aggregates are broken down producing an unfavourable structure. Alternatively, if ploughed too dry, huge clods result which are difficult to break down (Fig 8.12). Such soils are best ploughed in the autumn and then left over the winter so that the large clods are exposed to the action of frost which breaks them down. However, the soil in the spring will be wet. So you cannot take heavy machinery on to it since it will puddle. Furthermore, it will warm up slowly and this will delay planting. Drainage is therefore likely to be essential.

The need is to get these soils to granulate which will increase porosity. This is best achieved by incorporating large quantities of organic matter. One of the best ways to do this is to use grass crops as part of a crop rotation. The grass can be cut for hay and the remaining grass can then be ploughed into the soil. Liming and manuring may also help to break up a clay soil.

Sandy soils

These soils contain at least 70% sand and less than 15% clay. Sandy soils are easy to work and therefore are termed light. Their low organic and clay content means that the particles have little tendency to stick together to form clumps, so even when wet these soils are not sticky. Their plasticity, the ease with which they can be moulded, is low. Sandy soils have relatively large pores and so are well aerated and drain easily. Since the proportion of micropores, which hold the water when the soil is at field capacity, is low, sandy soils do not retain water well and are drought prone. However, they do warm up rapidly in the spring. Their low clay content means they are deficient in nutrients and so they are often described as being hungry soils.

Management of these droughty and often infertile soils involves increasing their granulation by adding organic matter which acts as a binding agent for the soil particles and increases its water- holding capacity. The addition of farm yard manure and the growth of grass are practices usually followed to improve the structural condition of sandy soils.

Loams

These are soils which consist of almost equal amounts of sand and silt and a little less clay. They have a medium porosity which means that, while there is ample space to provide oxygen for plant root cells, loams retain enough water for roots to absorb without becoming waterlogged. Their clay content means they have a reasonable reservoir of essential plant nutrients.

8.9 Table 8.5 shows the percentage water in different types of soil at field capacity and the wilting percentage.

Table 8.5 The percent water in different types of soil at field capacity and at the wilting percentage.

	Field capacity	Wilting percentage
fine sand	3.3	1.3
sandy loam	18.5`	10.0
silt loam	21.3	10.4
clay	28.0	14.5

(a) Assuming it is several hours since the last rain fell, calculate the percentage of available water in each soil.

(b) Explain why clay soils have a higher field capacity than sandy soils.

(c) Explain why sandy soils have a lower wilting percentage than clay soils.

8.10 (a) Why will a soil with abundant organic matter usually be better aerated than a clay soil?

(b) Clay soils are described as being cold because they warm up slowly in the spring. Why do such soils warm up so slowly and what are the problems this presents for farmers?

(c) To sow seeds earlier, gardeners are recommended to cover their land with polythene sheet or cloches. What is the physical basis of this practice?

SUMMARY ASSIGNMENT

1. Make sure you have definitions of soil texture, soil structure, humus, field capacity, permanent wilting percentage, clay soil, sandy soil, loam soil, organic soil.

2. (a) Compare and contrast the physical structures of clay, sandy and loam soils.

   (b) How will adding organic matter to clay and sandy soils improve their physical structure? Why will these improvements enhance plant growth?

3. Write an essay on the following: Soil is an ever-changing body of matter which responds to several influences. Name these influences and explain clearly how each affects the soil.

Chapter 9

SOIL MANAGEMENT

In Britain and other developed countries, farming is essentially a business. Like all businesses the farmer's aim is to make as large a profit as possible. To achieve this the farmer needs to achieve high yields of the crop(s) which provide the best economic return as efficiently and as cheaply as possible. To obtain high yields, farmers need to manipulate the environment of their crops. Given that farmers have to accept the constraints imposed by climate in regard to heat and light and to make do with the carbon dioxide in the air the three main areas of environmental management open to farmers are shown in Fig 9.1. Note that this diagram emphasises the interaction between the three components of the management triangle. For example, a crop growing in a fertile, well aerated soil and supplied with ample water is likely to be less susceptible to attack by a pest than one which is not. The sorts of things that farmers do – ploughing, spraying herbicides, adding fertilisers – do not therefore affect just one component of the management triangle. You must keep this in mind as you work through this chapter.

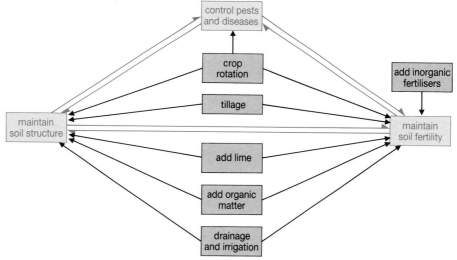

Fig 9.1 The farm management triangle. The farmer's objectives are set out in the red boxes. The way these are achieved are indicated in the grey boxes. Note how the factors interact. Thus improving soil structure will, by encouraging strong plant growth, reduce the impact of pests.

LEARNING OBJECTIVES

After completing the work in this chapter you will be able to:

1. explain the beneficial and detrimental effects of tillage and minimum cultivation on soils;

2. account for the use and effects of lime on agricultural soils;

3. discuss the use of inorganic and organic fertilisers;

4. describe how soil water levels can be managed through drainage and irrigation.

9.1 TILTH AND TILLAGE

A major aim of agricultural practice is to promote the formation of a good tilth which, simply defined, is the physical condition of the soil in relation to plant growth. Tilth depends on soil granule formation and stability, which, in turn, affects moisture content, degree of aeration, rate of water infiltration, drainage and field capacity of the soil.

Tillage and crop production

Tillage involves the mechanical manipulation of the soil to promote good tilth and hence high crop production. Tillage has three main aims.

1. To produce a suitable seed bed for sowing and seed germination.
2. To incorporate organic residues into the soil.
3. Pest control.

Primary tillage

In western agricultural practice, primary deep ploughing, using a mould board plough (Fig 9.2), is the traditional first step in soil tillage. The plough share (i.e. blade) slices down some 15–18 cm and the mould board behind it twists the slice over and throws it to one side. The whole action shatters the slice of soil producing a crude level of granulation which is improved by weathering. Under the action of frost and rain the original ploughed clods become granular lumps. These can be further broken down by secondary tillage.

The beneficial effects of ploughing include the following:

Pest control: Weeds are uprooted and turned into the soil; animal pests, e.g. wire worms and leather jackets, are brought to the soil surface exposing them to frost desiccation and predators.

Assisting soil granulation: This promotes aeration and drainage which is particularly important on clay soils.

Breaking up soil pans: These are layers of hard soil which have formed through soil compaction or leaching of nutrient salts.

Incorporating organic matter into the soil: This further improves soil structure, aids water retention and, ultimately, adds nutrients to the soil as the organic matter decomposes.

Bringing leached nutrient salts back to the surface layers: These nutrients are then available for uptake by plant roots.

Primary ploughing will not be an appropriate tillage operation on all soils or under all conditions. For example, it may be unnecessary on sandy or loamy soils or where shallow rooting crops are to be grown. Conversely, on clay soils or where deep rooting crops, like potatoes or sugar beet, are to be grown it is extremely useful. When soils are very thin, for example some of the calcareous ('chalky') soils of East Anglia, deep ploughing may be impossible or can lead to soil erosion. Finally, and perhaps most importantly, the economics of farming require intensive cultivation where one crop is planted immediately after the harvest of the preceding one (**double** or **continuous cropping**). This may leave no time for ploughing and some system of **minimal cultivation** (see below) will have to be used instead.

Secondary tillage

Conventionally, primary ploughing is followed by a number of secondary tillage operations such as harrowing, discing and rolling. These kill weeds, break up clods and prepare a suitable seed bed. The extent of these secondary tillage operations depends on the sort of seed beds required. For example, fine seed, like that of oil seed rape, requires a fine, firm seed bed while the seed bed for potatoes can be much coarser. After the crop has

(a)

(b)

Fig 9.2 **(a)** A traditional mould board plough, the most widely used of all primary tillage implements. Notice how the plough both cuts the soil and turns it over, burying weeds and previous crops.
(b) Secondary tillage. This tractor has a disc harrow mounted on the front which produces a fine tilth, allowing seeds to be drilled at the back. Notice that only one pass of the tractor is needed to complete both operations, essential on a clay soil like this if compaction is to be avoided. The wide double wheels help to spread the weight of the tractor, again to avoid compacting the soil.

SOIL MANAGEMENT

germinated further secondary tillage operations, e.g. hoeing, may be used to suppress weeds.

The benefits and costs of tillage

In the short term, tillage operations are generally beneficial.

- Crop residues break down more quickly if they are cut up and incorporated into the soil by tillage implements.
- Immediately after ploughing the soil is loosened and total pore space is increased, so providing better water infiltration, drainage and aeration of the soil.
- Weeds are suppressed.
- Fine seed beds can be prepared.

However, tillage leaves the soil surface bare and subject to wind and water erosion. Furthermore, the operations bring weed seeds to the surface, so stimulating their germination. The long-term effects of tillage, particularly ploughing, are generally undesirable since:

- mixing and stirring the soil increases the rate of oxidation of organic residues, so reducing soil organic matter and decreasing soil granule stability (Fig 9.3);
- tractors and other heavy implements compact the soil and break down the soil granules;
- continued ploughing produces a dense zone of compacted soil, a **plough pan**, immediately below the plough layer which inhibits root growth;
- high fuel and labour costs mean that extensive tillage operations are expensive.

All of these problems associated with conventional tillage have stimulated interest in, and the use of, agricultural systems which reduce the number of tillage operations.

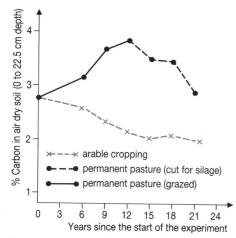

Fig 9.3 The changes in the organic content of ploughed (arable cropping) and unploughed (pasture) soils in the Highfield ley–arable experiment, Rothamsted, England 1949–72.

Graph legend:
×---× arable cropping
●--● permanent pasture (cut for silage)
●—● permanent pasture (grazed)

y-axis: % Carbon in air dry soil (0 to 22.5 cm depth)
x-axis: Years since the start of the experiment

Minimal cultivation

The development of effective herbicides in the late 1950s and early 1960s meant that weeds could be controlled without extensive tillage operations. This allowed the introduction of tillage sytems that minimised or even eliminated the use of the plough. These new tillage practices are collectively called **minimal cultivation** or **conservation tillage**. The area under minimal cultivation in Britain increased from less than 10 000 ha to almost 250 000 ha from 1969 to 1978.

Direct drilling is an extreme method of minimal cultivation. Here the seeds are planted directly into the soil, sometimes through the residues of the previous crop which are then allowed to decompose *in situ* (Fig 9.4) using a specially strengthened seed drill. Alternatively, stubble may be burnt off although this practice is now strictly controlled. Weeds are controlled using a contact herbicide, like paraquat, either before the seed is drilled or before it germinates. Rooting and drainage conditions are maintained by encouraging earthworm activity. Less extreme are the systems of minimal cultivation where the ground is just ploughed or the surface layers are tilled lightly using power rotovators that undercut any crop residues and weeds, stirring the soil to a depth of 6–10 cm but not inverting it.

A key element in all minimal cultivation operations is the use of herbicides to control weeds. In the spring, for example, a grass crop or a winter cover crop can be treated with herbicide before the new crop is planted through the grass. This stops the growth of the grass but leaves the

Fig 9.4 A direct drilling machine plants seeds through the remains of a previous crop without the need to plough first. The seed is contained in the large tank and drops down through the clear plastic tubes to the specially strengthened seed drills.

soil covered, so preventing erosion by spring rain and wind. In addition, the use of cover crops encourages water penetration, reduces evaporative water loss from the soil and reduces nutrient loss through leaching.

In general, then, the advantages of minimal cultivation are:

- reduced soil compaction and the development of a better soil structure;
- reduced land preparation time thus permitting continuous cropping;
- reduced soil erosion since the land is left with a covering of crop residues;
- reduced annual weed germination as the seed bank (see Section 10.2) is disturbed less;
- reduced fuel and labour costs. Conventional tillage requires the passage of a tractor, towing heavy implements, perhaps five to six times across a field to prepare the seed bed. Minimal cultivation may require just one pass.

On the negative side, minimal cultivation:

- is not appropriate for all soils, especially heavy, poorly drained clay soils where yields are reduced.
- increases the use of herbicides which are both expensive and potentially dangerous. Increased herbicide costs must be offset against reduced fuel costs.
- reduces soil aeration. This leads to a reduction in the oxidation of organic residues and possibly nitrogen deficiency. This can necessitate the use of larger amounts of inorganic nitrogen fertiliser to maintain yields.
- can promote soil surface acidification since fertilisers are usually placed on the soil surface and the acidifying effects of applied nitrogen (see later) on the upper soil layers may be detrimental. This can be overcome by occasional ploughing to mix the soil layers.

QUESTIONS

9.1 Why will ploughing both cure and cause weed problems?

9.2 Under what circumstances will ploughing be (a) beneficial (b) detrimental to the structure of a clay soil?

9.3 Directly drilled soils show dramatic increases in the number of earthworms compared to ploughed soils.
 (a) Explain this observation.
 (b) What are the likely beneficial effects of an increase in the number of earthworms?
 (c) Why will minimal cultivation reduce soil erosion?

9.4 Directly drilled soils tend to warm up more slowly in the spring than ploughed ones.
 (a) Explain this observation.
 (b) What are the implications of this observation for farmers growing (i) winter (ii) spring sown cereals?

9.2 LIMING

The maintenance of soil fertility in humid agroclimatic zones, like Britain, depends on the careful use of 'agricultural lime', a term which covers three different sorts of compound (Table 9.1). When lime is added to the soil, two general changes occur to it. Firstly, the calcium and magnesium compounds in the lime dissolve in the presence of carbon dioxide. Secondly, these dissolved compounds react with soil colloids, clays and humus particles, displacing H^+ ions by cation exchange (see Fig 8.4) and so raising the soil pH. Between applications of lime, magnesium and calcium are lost by erosion, crop removal and leaching. Consequently, the pH of the soil

Table 9.1 The principal types of agricultural lime. (Note that the oxide and hydroxide of lime are made by heating limestone (oxide) or by heating and then adding water (hydroxide). In addition to calcium, limes contain at least some magnesium.)

Compound	Chemical formula	Other names
oxide of lime	CaO	quick lime, burnt lime
hydroxide of lime	$CaOH_2$	hydrate, slaked lime
limestones	$[CaCO_3]_2$	calcite
	$[CaMg(CO_3)_2]$	dolomite

gradually falls and eventually another application of lime will be needed. Other factors which increase soil acidity include the following:

- The continued application of nitrogen fertilisers.
- Waterlogging which by reducing aerobic bacterial metabolism can lead to an increase in the concentration of humic acids in the soil and the formation of hydrogen sulphide by anaerobic bacteria. Hydrogen sulphide can then be converted into sulphuric acid.
- Acid rain.

Effects of lime on the soil

Lime changes its form in many complex ways after it has been added to soil but we can group the effects of liming on soil under three headings – physical, chemical and biological.

Physical effects

Lime promotes clumping (**flocculation**) of soil particles, thus improving soil structure. This effect is particularly noticeable on clay soils. The mechanisms involved are indirect. For example, liming encourages the activity of soil microorganisms, which promotes the decomposition of organic material and the formation of humus. The production and maintenance of high humus levels in the soil encourages granulation, so improving soil structure.

Chemical effects

Most plants grow best at about pH 6.5 so raising an acid soil's pH to this level by liming will increase crop yields. The main reason for this enhanced growth is not due to a reduction in acidity *per se* but to:

- an increase in the availability and plant uptake of elements such as molybdenum, phosphorous, calcium and magnesium;
- a reduction in the concentration of iron, aluminium and manganese which, under very acid conditions, are likely to be present at toxic levels (see Fig 8.5).

In addition, lime is the main source of the essential plant nutrients calcium and magnesium.

Biological effects

Liming encourages the activity of soil microorganisms like bacteria and fungi. This increased activity favours

- the formation of humus;
- the release of nitrates and sulphates from decomposing organic matter;
- the activity of nitrogen-fixing bacteria which live in the root nodules of plants like clover;
- the metabolism and elimination of potentially toxic organic compounds in the soil.

(a)   **(b)**

Fig 9.5 **(a)** Common scab (*Streptomyces scabies*) symptoms on potato tubers. Whilst this does not render the tubers inedible, it does make them less commercially valuable – people don't buy potatoes with disfigured skins.

(b) The effect of club root (*Plasodiophora brassicae*) is to deform the root growth of all types of brassicas, like cabbages and brussel sprouts, which stunts the growth of the plant. This soil transmitted fungus is practically impossible to eradicate from infected soil.

Unfortunately, liming may also stimulate the growth of undesirable microorganisms, for example the actinomycetes fungus which causes potato scab (Fig 9.5(a)), although the growth of some plant pathogens, for example club root of brassicas (Fig 9.5(b)), are inhibited by liming.

The amount of lime applied is affected by a number of factors, including:

- soil pH, texture, structure and organic matter content;
- the crops to be grown;
- the type and fineness of the lime to be applied;
- economic returns in relation to the cost of the lime.

The cost of liming will depend on the type of lime used (the more processed it is the more expensive it is) the fineness (the finer it is the more expensive it is) and transportation and application costs. Generally, it is seldom economical to apply at more than 7–9 tonnes ha^{-1} of finely ground lime to a mineral soil unless it is very acid. Lime is best applied to ploughed land and then worked into the seed bed by secondary tillage operations. This may be impractical, even undesirable if it causes soil compaction, so it is often more convenient, and generally just as effective, to apply lime to the surface of the ground and then plough it in.

It does not really matter at what time of the year the lime is applied but in a rotational system of cropping it makes sense to apply it with or ahead of the crop that gives the best response, usually the legume in the rotation.

QUESTIONS

9.5 **(a)** How would you determine whether it was necessary to apply lime to soil?

(b) What problems might arise if you over-limed a soil? (**HINT:** look at Fig 8.5.)

(c) Lime is often applied to the surface of a grass crop in the autumn which is then ploughed in. Explain this practice.

9.6 On allotments where vegetables are grown a typical four-year rotation would be

> potatoes → brassicas → legumes → root crops

Explain when and how much (heavy or light) lime you would apply in such a rotation.

9.3 FERTILISERS

Any inorganic salt, like ammonium phosphate, or organic substance, like manure or sewage sludge, which is applied to soils to increase yields by supplying plant nutrients is a fertiliser. The use of organic materials dates back hundreds of years but mineral salts have only been used extensively in the last 100 years. Nonetheless, it is these inorganic fertilisers which are now used most in British agriculture and so we will concentrate on them in this section. Organic supplements, like manure, will be dealt with in the next section.

Fertiliser use

From 1939 to 1981 the amount of inorganic fertiliser used in the United Kingdom increased by 550%. Much of this increase was due to the dramatic fall in the price of nitrogen fertilisers, which make up 73% of the total nitrogen, phosphorous and potassium fertilisers used, as a result of the development of the Haber process which converts atmospheric nitrogen into ammonia. Thus, in 1986–87, 1.67 million tonnes of nitrogen fertilisers were spread on to land in the United Kingdom compared with 60 000 tonnes in the 1930s. This huge increase in fertiliser application is one of the main reasons, along with improved pest control, for the enormous increase in yield of arable crops, particularly cereals.

However, we do pay a price for the cheaper food which the use of nitrogen fertilisers has produced. Plant roots can only absorb nitrogen in the form of nitrate or ammonium ions. Neither of these ions are adsorbed on to soil particles but remain in the soil solution. As a result they are easily lost from the soil by leaching. The nitrate and ammonium ions drain down through the soil and rock either entering the ground water, where they can contaminate drinking water, or running off into nearby lakes, reservoirs and rivers which can lead to cultural eutrophication (see Chapter 13). In addition, the European Commission limit of 50 mg nitrate per dm^3 for drinking water is consistently passed in some areas of England, for example in East Anglia where there is intensive cereal cultivation.

Types of fertiliser

Plants need some 14 essential mineral nutrients to grow. Of these, six are classified as major – calcium, magnesium, nitrogen, phosphorous, potassium and sulphur – since they are needed in large quantities by plants (Table 9.2). Calcium and magnesium are supplied by lime. The sulphur needs of plants are largely met by natural sources, e.g. the weathering of rock and sulphate dissolved in rain water. Nitrogen (N), phosphorous (P) and potassium (K) can be applied separately as so-called **straight fertilisers**, which deliver just one nutrient, or as a **combination fertiliser** (Fig 9.6). Inorganic fertilisers come in both liquid and solid forms. Solid fertilisers make up about 95% of total applications in the United Kingdom and are commonly provided in the form of granules or pellets for spreading by machine (Fig 9.7).

Fig 9.6 A combined NPK fertiliser. The composition of manufactured fertilisers is controlled by EC regulations.

Fig 9.7 Granulated fertiliser being spread onto a young cereal crop. The machine has a spinning disc which throws the pellets of fertiliser in a broad band behind the machine. The pellets make the fertiliser particularly easy to handle and store, as well as providing a means of delivering the exact amount required to the crop.

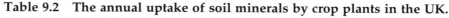

Table 9.2 The annual uptake of soil minerals by crop plants in the UK.

Mineral	Amount removed by crops (kg ha⁻¹ yr⁻¹)
nitrogen	100
potassium	100
calcium	50
sulphur	30
phosphorous	15
magnesium	15

Principles of fertiliser application

To get the most out of a fertiliser it has to be supplied

- in the right form;
- in the correct amount;
- at the right time;
- in the right place.

Form and amount of fertiliser

Until recently it was thought that there was little difference on crop yield between the forms of inorganic fertiliser. The main criterion for use was economic: use the fertiliser which gives you the most for your money. This involves buying the fertiliser which gives the greatest amount of nutrient per unit price, not the greatest amount of fertiliser per unit price. However, recent evidence suggests that crops do respond differently to different fertilisers. Thus grass yields, for example, seem to be greatest when nitrate is applied as ammonium nitrate.

Table 9.3   **Fertiliser recommendations for selected crops.**

Crop	Fertiliser recommendations (kg ha^{-1})		
	N	P	K
winter wheat			
(after cereals)	150–200	50–100	50–100
(after roots)	50–100	40–60	0–60
spring barley			
(after cereals)	125–150	50–100	50–100
(after roots)	40–75	40–60	0–60
sugar beet	125–150	65–125	150–200
potatoes	120–200	150–220	150–250
oilseed rape	120–200	60–80	60–80
maize	100–150	60–90	60–90
mown grass	200–400	80–100	60–200
grazed grass	150–350	50–60	60–100

The amount of fertiliser applied will depend on the type of crop (Table 9.3) and its position within the crop rotation. Different crops have quite different nutrient demands. For example, barley has large nutrient demands, severely depleting N, P and K reserves in the soil. By contrast, brassicas, like cabbage, are far less demanding. Cereal crops only recover about one-third of the nitrogen applied to soil. Root crops are slightly more efficient, recovering about a half with grass recovering about three quarters. The rest is lost either by leaching or by denitrification. **Denitrification** involves the microbial conversion of nitrate to gaseous nitrogen. The rate of this process is increased by the anaerobic conditions produced by waterlogging of the soil. By using a suitable crop rotation, farmers can ensure that they use all the nutrients they apply to the soil as efficiently as possible.

Timing of fertiliser application

This is vital since you must supply nutrients when the crop most needs them otherwise growth will be suppressed, leading to reduced yields and possible disease problems. Fertilisers must therefore be applied when the nutrient demands of the crop are at their greatest. Nutrient uptake is usually greatest during early growth, reaching a maximum in cereals during **tillering** (the development of side shoots from the main body of a cereal plant each of which eventually produces a flower spike) and flowering stages, and declines during the ripening phase as nutrients are moved from the leaves and stalks to the developing grain. However, you also need to take account of the mobility of the fertiliser, where the fertiliser is placed in the soil, and the prevailing weather and soil conditions.

Fertilisers can be applied either before, during or after sowing. The major consideration in selecting the timing of fertiliser application is the type of nutrient. Thus nitrogen, which can only be absorbed by plants as ammonium or nitrate ions, both of which are highly soluble and highly mobile, is rarely applied to winter wheat when it is sown in the autumn because of the possibility of nutrient loss by leaching and denitrification in the wet winter months. Consequently, nitrogen is applied to winter cereals in the spring. While an early spring dressing in, say, March or April, when soil temperatures are high enough for growth to restart, may seem best, to promote early growth, other factors may militate against this. For example, heavy spring rains may still cause considerable losses of fertiliser by leaching while taking heavy machinery on to wet soil may damage the soil structure. Dressing late in the spring (late April–May) avoids these problems. By contrast, fertiliser application to spring-sown cereals, like spring barley, is usually by either a combined drill, which applies both fertiliser and plants the seed, or by broadcasting (i.e. scattering) the fertiliser prior to planting.

The application of phosphorous fertilisers is somewhat different. Phosphate is rapidly fixed in the soil, as it is absorbed on to exchange sites on the surface of soil colloids, rendering it practically immobile. This means that phosphate is not easily leached from the soil and there is a marked carry-over effect from previous years. For this reason it is rarely necessary to apply phosphorous fertilisers to the soil every year. The levels of phosphate in the soil can be satisfactorily maintained by ploughing in phosphate fertilisers during the autumn, perhaps every three to five years.

ANALYSIS

Fertilisers and edaphic factors
This exercise requires you to use analytical thinking skills.

Any factor which limits plant growth will reduce fertiliser efficiency and hence reduce the crop's response. It is only when other factors, such as soil aeration, water supply, light intensity and so on, are not limiting (remember the law of limiting factors) that fertilisers increase yield. Use this knowledge to explain the following observations.

(a) Yields of barley are greater on land that has been previously used to grow grass.
(b) In 1976, a drought year in Britain, winter wheat yields on freely draining limestone soils with N added at 60 kg ha^{-1} were 3 tonnes ha^{-1}. In 1977, when rainfall returned to its usual levels yields increased to 4.5 tonnes ha^{-1} even though N was added at the same rate.
(c) Boulder clay soils yielded 55% more wheat in 1976 than limestone soils even though N was applied at the same rate on both soils.
(d) Farmers apply small but regular applications of N to grassland which is grazed continuously throughout the year.

Disadvantages and advantages of inorganic fertilisers

In addition to the problems associated with nutrient run-off, which are dealt with in more detail in Chapter 13, other disadvantages of using inorganic fertilisers include the following.

Plant damage: Since nitrogen fertilisers are so soluble they can damage plants if they are placed too close to the roots or they are brought close to the roots by the upward movement of the soil solution by capillary action. This damage, often called 'burning' or 'scorch', results from the high osmotic potentials exerted by the soil solution caused by the high concentrations of nitrate it contains.

Soil acidification: Complete fertilisers, unless specially treated, tend to develop an acid residue in the soil. This results mainly from the nitrification of ammonium compounds used as nitrogen carriers in the fertiliser since, on oxidation, the ammonium compound increases acidity:

$$NH_4^+ + 2O_2 \rightarrow 2H^+ + NO_3^- + H_2O$$

Cost: Inorganic fertilisers are expensive.

However, despite their expense and the problems associated with their misuse, inorganic fertilisers do have advantages if used sensibly. These include:

- their concentrated nature which makes for ease of storage and application;
- a known nutrient content – you know what you are putting on the soil enabling you to match nutrient supply and demand;
- a high nutrient content – you get a lot for your money;
- they can be applied using light machinery, so reducing soil compaction;
- they can be applied evenly to a field.

You should compare these with the advantages and disadvantages of using organic sources of fertilisers (see Section 9.4).

ANALYSIS

Economics and fertiliser application
This is a data-analysis exercise.

In carrying out any farming operation, for example fertiliser or pesticide application, the farmer must take economics into consideration. In particular, it is important that the value of increased yield resulting from the operation (called the **yield return**) must exceed the cost of the operation. Costs include the expense of the raw materials, and the costs of machinery, fuel and labour. This important management point is highlighted in this exercise.

Table 9.4 The yield response of a cereal crop to different levels of nitrogen fertiliser application.

N applied (kg ha⁻¹)	Yield (t ha⁻¹)
0	2.8
50	3.3
100	4.2
150	4.7
200	4.8

Table 9.4 shows the yield response of a cereal crop to different levels of nitrogen fertiliser application.

(a) Draw a graph of the data and describe the relationship.
(b) Assume that the value of the crop is £100 per tonne and the cost of the fertiliser is 50p per kg. Calculate the benefit–cost ratio which is given by the value of increased yield/cost of fertiliser used.
(c) Do benefits exceed costs at all rates of fertiliser application?
(d) How does the benefit–cost ratio change as the rate of fertiliser application increases?
(e) Would it be worthwhile applying N at rates exceeding 200 kg ha⁻¹?
(f) Which level of N application gives the maximum net yield return (value of increased yield – cost of fertiliser application)?

QUESTIONS

9.7 Intensive cereal cultivation requires the use of large amounts of nitrogen fertiliser to maintain yields. Given that cereals only recover about one-third of the nitrogen applied what does this suggest about **(a)** the need for applying lime to the soil growing these crops

---

(b) the frequency of application of fertiliser **(c)** the ultimate destination of the lost fertiliser? **(d)** Wheat flour used in bread making needs to have a high protein content. Why would it be a good idea to apply extra nitrogen fertiliser to bread wheat as the ears begin to develop?

9.8 List all the disadvantages of using inorganic fertilisers.

9.9 **(a)** Suggest why potassium levels in soils at Rothamsted not treated with potassium fertilisers since 1901 still have potassium levels well above those of totally untreated soils.

(b) What is the implication of this for applying potassium fertilisers to the soil?

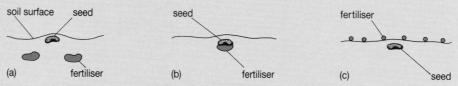

Fig 9.8   Methods of applying fertiliser. The method you employ depends on the type of fertiliser.
(a) Side band application below seed.
(b) In contact with the seed.
(c) Broadcast application on the surface.

9.10 Fig 9.8 shows different ways of applying fertilisers.
(a) Why will placing the fertiliser in contact with the seed using a combine drill not be a suitable way to apply a nitrogen fertiliser like urea but ideal for phosphate?
(b) 'Burning' of young plants can occur when their roots come into contact with soil solutions containing high concentrations of nitrates. Plants are particularly prone to this sort of damage when rain, immediately after planting, is followed by a long dry spell. Explain these observations.

9.4 MANURING

In this section, we will consider three sorts of organic matter applied to the soil in Britain – green manure and crop residues, farm yard manure and sewage sludge.

Green manure

This involves ploughing in a crop such as grass, rye or buck wheat; ploughing in residues from a previous crop; or leaving the previous crop on the surface and direct drilling through it. In all cases the green manure adds organic matter to the soil which slowly releases plant nutrients as it decays. In addition, a green manure can act as a cover crop during its growth. This prevents erosion and plant nutrients, which would otherwise be leached away from bare soil, are retained in the crop.

All manures do present one problem for the farmer. Manures contain lots of carbon but little nitrogen, i.e. they have a large C/N ratio. So adding a manure to the soil effectively adds lots of carbon but little nitrogen. As the manure is broken down and the carbon-containing compounds respired to produce carbon dioxide the number of microorganisms which are responsible for the decay increases. These microbes cannot live on carbon alone. To grow and reproduce they need the same sorts of nutrient as plants do. In particular, the growing microbial biomass mops up the available soil nitrogen, e.g. nitrates and ammonium ions, which can lead to a short-term deficiency of these substances in the soil. As the microbes start to die and decompose themselves these nutrients will be released back into the soil and so become available again, but this may take several years. In the meantime, supplementary fertilisers will be needed.

Farmyard manure

Farmyard manure (FYM) has been used for centuries as a means of improving soil fertility and structure. FYM consists of the urine and faeces of animals usually combined with bedding material like straw. Biologically, manure has many attributes. It supplies a wide variety of nutrients plus organic matter, which improves soil structure, water retention and aeration and increases the cation exchange capacity of the soil. It certainly has beneficial and long-term effects on plant growth and the soil (Table 9.5).

Table 9.5 Yields of crops (t ha^{-1}) under long-term applications of farmyard manures and inorganic fertiliser.

Crop	Control (no fertiliser)	FYM only	NPK only	FYM as % of NPK
wheat	2.08	3.50	3.11	112
barley	1.03	2.03	2.26	90
sugar beet	3.80	15.60	15.60	100
mangolds	3.80	22.30	30.90	72

However, FYM does have a number of disadvantages as a fertiliser, including the following.

- Low nutrient content (Table 9.6): This means that you have to spread a lot of it to get the same effect as an inorganic fertiliser. This means high labour and fuel costs, which can make FYM, as a fertiliser, economically uncompetitive. The heavy machinery used also causes soil compaction (Fig 9.9).

Table 9.6 The chemical composition of farmyard manures and slurries. Values are % composition.

Manure	N (%)	P (%)	K (%)
cattle manure	0.6	0.1	0.5
pig slurry	0.2	0.1	0.2
chicken manure	1.5	0.5	0.6
sewage sludge	1.0	0.3	0.2

- Variable nutrient content depending on animal species, diet and so on (see Table 9.6). This means that you do not know what you are adding.

- Nutrient imbalance: FYM contains little phosphorous so you have to supplement this nutrient.

- Slow release of nutrients: Although this may be beneficial in the long run it does not help a farmer to meet the short-term nutrient needs of crops, for example when they need a quick shot of nitrogen.

- Pest control: FYM may contain viable weed seeds and the spores of plant pathogens.

- Uneven distribution of nutrients: Muck spreading machinery provides an uneven distribution of the manure and hence its nutrients (Fig 9.9).

- Handling difficulties: FYM is bulky and heavy so many trips will need to be made to the muck pile.

- Transport problems: FYM may not be available locally in sufficient quantities so it has to be transported in from elsewhere. This again raises costs.

Fig 9.9 Muck spreading onto stubble prior to ploughing in the late autumn. Note how delivery of this valuable material is poorly controlled compared to the pellets we saw in Fig 9.7. The heavy machinery needed to spread muck also means that you cannot apply it to growing crops like wheat or on wet soils where compaction would occur.

SOIL MANAGEMENT

- Wide C/N ratio: This means that it will pay the farmer to store the manure before spreading it. During storage it will start to rot, so reducing the amount of carbon. At the same time nutrients will be released from the organic compounds in a soluble form. However, storage presents its own set of problems (see Chapter 13).

Despite these drawbacks FYM, as with all manures, is a valuable addition to the soil because the organic matter it contains ultimately forms humus, that almost magical substance which does so much for soil structure.

Sewage sludge

This material is formed as a by-product in sewage works. As a fertiliser it has many of the drawbacks associated with FYM although it is very cheap and plentiful. In addition, sewage sludge may contain high concentrations of heavy metals like lead, cadmium, zinc, copper and nickel which may prove toxic to plants. There are also worries that if applied continually to pasture the levels of these heavy metals could build up in the grass and then be passed on to people via cows and their milk.

ANALYSIS

Crop rotation

This is a comprehension exercise in which you will need to use all of the knowledge you have gained so far in this theme.

Traditionally, crop rotation has been an important method for maintaining soil structure and fertility. A typical four-year rotation, in common use in the last century, is shown in Table 9.7.

Table 9.7   The Norfolk four-year rotation.

Year	Crop	Use
1	turnips or swedes	winter feed to sheep who feed in fields
2	spring barley	cash crop
3	red clover	grazed in spring and summer
4	winter wheat	cash crop

(a) Why is spring barley planted in the second year?
(b) What is the function of the red clover in the third year?
(c) Prior to planting winter wheat the clover would be ploughed in. What would the beneficial effects of this be on soil structure and soil fertility?
(d) Discuss the importance of animals in this rotation.
(e) How would the use of winter root vegetables like turnips help to preserve soil nutrient levels?
(f) Suggest why the use of such crop rotation has been replaced by crop rotations like those shown in Table 9.8?

Table 9.8   Typical arable rotations in Britain.

Year	Crop rotations			
	A	B	C	D
1	early potatoes/kale	potatoes	wheat	sugar beet
2	wheat	wheat	barley	barley
3	sugar beet	wheat	barley	barley
4	barley	peas	barley	vegetables
5	barley	barley	pea	wheat
6		barley	seeds	

(g) Account for the development of three crop rotations and systems of monoculture which just grow one crop, usually cereals, on one piece of land for many years, on the rich agricultural lands of East Anglia and the fens. What problems might you predict as a result of these changes in agricultural practice?

9.11 (a) Compare and contrast the benefits of using organic and inorganic fertilisers on soil fertility.
(b) Why is farmyard manure usually stored and left to decompose before being applied to the soil?

9.12 Disposal of sewage sludge by spraying it on farm land is becoming increasingly popular. Such sludge may be contaminated with potentially toxic heavy metals like lead, cadmium, copper and zinc. Why should such contaminated sludge not be applied to acid soils?

9.5 SOIL WATER MANAGEMENT

Too much or too little water can cause reductions in crop growth. Poorly drained soil may become waterlogged and the resulting poor aeration inhibits the uptake of nutrients by plant roots, and may lead to anaerobic conditions when toxic compounds form in the soil and increase the activities of denitrifying bacteria, so increasing losses of nitrogen from the soil. In addition, such soils are slow to warm up in the spring which will inhibit germination and root growth. Finally, using heavy machinery on a waterlogged soil has disastrous effects on soil structure, producing a slimy, amorphous mess. The only way to deal with this problem is to increase the drainage.

Soil drainage

The methods of land drainage include surface drainage, mole drainage and underground drainage.

Surface drains or ditches are cheap to make and carry large amounts of water. The negative aspects include the high costs of maintenance, interference with agricultural operations (e.g. ploughing) and consumption of valuable agricultural land. Surface drains work best when they are constructed across a slope and across the direction of cultivation. This enables the ditches to intercept the water as it moves down the slope.

Mole drainage is created by pulling a pointed cylindrical plug 7–10 cm in diameter through the soil (Fig 9.10(a)), forming a channel with a compressed wall which acts as a drain. These tunnels remain open for many years.

Underground drains, although more expensive to construct, provide the most efficient means of draining land, provided they are kept free of obstructions. Three methods are commonly used.

1. A **perforated plastic pipe** can be laid underground using special machinery. Water moves into the pipe through the perforations and can then be led to an open outlet ditch.
2. **Tile drains** are made of individual clay pipes 30–40 cm long. These are laid at the bottom of the open ditch and then covered with a thin layer of straw, manure or gravel. The ditch is then filled in with soil (Fig 9.10(b)).
3. **Subsoiling** involves using a special, deep plough share. This lifts and shatters the subsoil below normal plough depth, so creating cracks and spaces which help drainage.

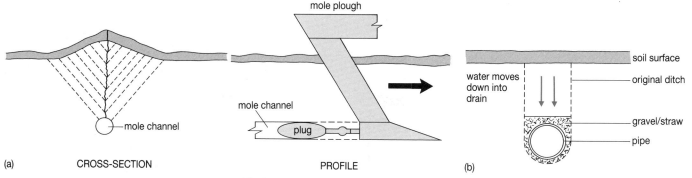

Fig 9.10 **(a)** A mole plough makes a channel through which drainage water can then move.
(b) A tile drain, although more expensive than mole ploughing, provides a more permanent means of drainage. The problem here is that the drains silt up.

Table 9.9 **The effects of agricultural drainage on crop yields. Spacing refers to the distance between drains.**

Location and crop	Spacing (m) and yield (t ha^{-1})			
Drayton – winter wheat	control 3.69	60 m 4.42	30 m 4.29	15 m 4.45
Cambridge – winter wheat	control 2.51	27 m 3.56	10 m 3.98	5 m 3.54

The effects of draining land on crop productivity can be dramatic (Table 9.9). In particular, drainage:

- reduces 'heaving' – the alternate expansion and contraction of the soil due to the freezing and thawing of soil water – which can dislodge the roots of crops, destroy roadways and break down soil granules, so affecting soil structure;
- allows the soil to warm up rapidly in the spring;
- increases soil aeration;
- allows farmers to work the land two to three weeks earlier in the spring.

The need for drainage depends on soil type. Clay soils and the rich peatlands of, for example, the mosses in southwest Lancashire and the fens of East Anglia, could not be farmed without drainage. In contrast, chalky soils, for example those of the downs in Southern England, are free draining and do not need such extensive drainage.

Mulching and irrigation

At certain times of the year, even in countries like Britain, there may be a shortage of water in the soil. Such **soil water deficits** (i.e. the amount of water needed to saturate a soil) are particularly damaging during flower formation and fertilisation, when insufficient water may cause the flower to fall off (e.g. tomato) or not to 'set' properly (e.g. cucumber), and germination, when seedling emergence will be delayed, so shortening the growing season. Prolonged water shortage at any time will reduce transpiration and hence reduce nutrient uptake, leading to a loss of yield. This problem will increase in importance as the crop canopy develops; for example, leafy plants like lettuces require large amounts of water as they mature. In addition, any crop which produces watery fruits, for example tomatoes and cucumbers, will require large amounts of water during fruit development. The soil water deficit will increase in the summer and reduce to zero in the winter in temperate climates. This **soil moisture deficit** can be managed in two ways: mulching and irrigation.

Mulching involves spreading crop residues, manure, peat, bark or even plastic sheeting on the surface of the soil, so reducing water loss by evaporation. Although impractical on a large scale it is widely used in horticulture and specialised areas like hop growing.

Irrigation is the addition of water to the soil, either in bulk or as a spray. We usually associate irrigation with hot, arid countries but crops growing in the drier eastern and southeastern parts of Britain can benefit from irrigation in most summers.

Irrigation in many areas is needed because of the inadequacy of the water supply in terms of rainfall at certain times of the year. Then:

Irrigation need = losses from evaporation + losses from drainage — rainfall

Clearly, then, a knowledge of water loss from the soil and precipitation will enable a farmer to supply irrigation water only when needed and thus avoid wasting water and energy. Such information can be provided by meteorological stations, or by monitoring water in the soil, and of course by experience.

Water can be applied to crops by three methods of irrigation: surface methods, overhead methods and subsurface methods. Of these, **surface methods** are the most widely used since they involve low cost technology and are easy to use over large areas. In temperate agriculture systems. **furrow irrigation** and **corrugation irrigation** are the most common methods of surface irrigation. In furrow irrigation (Fig 9.11) water runs down specially constructed parallel furrows. It is widely used with row crops like vegetables. Corrugation irrigation uses shallower furrows which are packed closer together than in furrow irrigation.

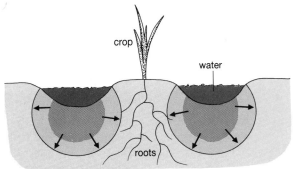

Fig 9.11 Furrow irrigation. Note how the water penetrates the soil close to the plant's roots.

Overhead application usually involves some sort of spray or sprinkler system. Such sprays may be static, rotating or may travel backwards and forwards. They are all easy to control so that water application can be easily controlled. However, it is a relatively expensive method and may damage soil structure by causing capping or soil erosion.

Subsurface irrigation is by far the best method. It is usually achieved by blocking underground drainage channels. Such a system minimises evaporative water losses, does not cause soil erosion or structural damage, and supplies water directly to the roots. In addition, it encourages deep rooting, thereby ensuring better nutrient supply.

Two problems associated with irrigation are **salinisation** and water logging. As irrigation water flows over and through the ground it dissolves salts, increasing the salinity of the water. Much of the water in this saline solution is lost to the atmosphere by evaporation, leaving salts, like sodium chloride, behind in the top soil, reducing crop growth. An estimated one-third of the world's irrigated land is now affected by salinisation. Since it is

SOIL MANAGEMENT

virtually impossible to irrigate sufficiently without applying excess water, or applying water continuously to avoid the problems of salinisation, good, well-maintained drainage sytems are essential if waterlogging of irrigated soils is to be avoided.

QUESTIONS

9.13 (a) Why will drainage increase the loss of nitrate fertilisers from soils?

(b) Drainage helps to prevent salinisation. Explain this observation.

(c) Why will drainage improve seed germination and crop growth in the spring?

9.14 (a) When would you apply irrigation to soil?

(b) Why is corrugation irrigation a better technique than furrow irrigation on steep slopes?

(c) Why are plants supplied with extra nitrogen fertiliser better able to withstand moisture stress?

SUMMARY ASSIGNMENT

1. Compare and contrast the processes involved in traditional cultivation with those of minimal cultivation.

2. **(a)** Distinguish between the terms soil structure and texture.

   (b) Describe the methods used by farmers to develop and maintain a good soil structure, indicating particularly the time scale involved.

   (c) How and why are improvements in soil structure likely to influence crop yields?

3. **(a)** Name one natural process which causes an increase in the acidity of the soil.

   (b) Give one reason why a low soil pH may result in a low crop yield.

   (c) How may this acidity be corrected by the farmer?

4. **(a)** Why is lime sometimes applied to soils? Explain fully both the direct and indirect effects of liming agricultural land.

   (b) List the liming materials commonly used and mention factors which determine the rates and timing of their application.

5. Discuss how fertiliser application will be affected by (i) the type of crop (ii) climatic conditions (iii) soil type.

6. Keep a note of your answer to Question 9.11 to remind you of the advantages and disadvantages of different types of fertiliser.

7. **(a)** How is crop growth affected by a shortage of water in the soil?

   (b) Discuss the factors which lead to soil moisture deficits.

   (c) Describe some of the methods used to irrigate crops in the UK. Using examples suggest when, in the sequence of crop growth, irrigation is likely to be most beneficial and give reasons for your suggestions.

Chapter 10

PESTS – BIOTIC FACTORS AFFECTING CROP PRODUCTIVITY

Fig 10.1 Pests of stored grain, like this grain weevil, *Sitophilus granarius*, can be devastating. The weevil, and its larva, burrows into the grain leaving just a hollow husk.

In addition to physical (abiotic) factors, light, temperature and so on, which place limits on crop productivity, Fig 7.1 shows that biological (biotic) factors also limit crop yields in three ways.

1. Other plants use essential resources, water, nutrients and light, which our crop plant could have used. If these essential resources are in short supply this may lead to a reduction in crop yield through the processes of **intra-** and **inter-specific** competition.
2. The plant, or more usually parts of it, are consumed by animals which can reduce the photosynthetic area or use up products of photosynthesis which could have been used by the plant for growth or eaten by humans.
3. The plant can contract a disease which impairs its metabolism and so slows its growth, or which renders the crop commercially less valuable.

In addition, there can be losses during storage (Fig 10.1). These biotic factors interact with each other, and with abiotic factors, to produce their effect on crop yield. Thus a plant which is being attacked by aphids, or which is stressed by a lack of water, may succumb more easily to a fungus infection or suffer a loss of competitive ability. In this chapter, we will concentrate on the types of organism which become pests. In the next chapter, we will look at how they are controlled.

LEARNING OBJECTIVES

After completing the work in this chapter you will be able to:

1. explain why and how an organism becomes a pest;
2. distinguish between and give examples of winter annual, summer annual and perennial weeds;
3. describe the effects of insect pests on crop plants;
4. describe the effects and mode of transmission of viral and fungal diseases of plants.

10.1 WHAT IS A PEST?

Any organism causing harm or damage to people, their crops or possessions is a pest. Agriculturally, an organism becomes an **economic pest** when the loss in crop yield exceeds certain proportions, usually 5–10% of total yield. Clearly, it only makes economic sense to control a pest when it is actually causing economic damage, i.e. when the costs of control are less than the economic damage being done to the crop. In other words, pest is not a biological concept it is an economic one.

The lowest population density of a pest that will cause economic damage is the **economic injury level (EIL)**, and will vary between crops, seasons

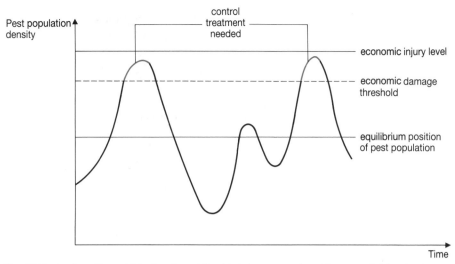

Fig 10.2 A schematic model to show the relationship between a pest and its economic injury level, economic damage threshold and general equilibrium levels. Note that an organism only becomes a pest when it exceeds its economic damage threshold.

and areas. It is of basic agricultural importance that it is known for all major crops in an area. Now you might think that control should only be applied when the pest exceeds the EIL. However, pest populations grow rapidly and, since it will take some time for the pest control measure to take effect, action needs to be taken before the pest reaches the EIL. So in practical terms EIL is not as important as the **economic damage threshold** (Fig 10.2). This is defined as the population density of an increasing pest population at which control measures should be started to prevent the population from reaching the EIL.

Using the model shown in Fig 10.2 we can recognise four types of pest.

1. **Non-economic pests** whose general equilibrium position and highest fluctuations are below the economic damage threshold, e.g. the aphid *Aphis medicaginis* on alfalfa in the United States.
2. **Occasional pests** whose general equilibrium position is below the economic damage threshold but whose highest population fluctuations exceed the economic damage threshold, e.g. the millipede *Brachydesmus superus* which occasionally damages young seedlings in Britain.
3. **Perennial pests** whose general equilibrium position is below the economic damage threshold but whose population fluctuations frequently exceed the economic damage threshold, e.g. the fungus barley powdery mildew *Erysiphe graminis hordei*.
4. **Serious pests** whose general equilibrium position is above the economic damage threshold. For example, in southwest Lancashire two species of weed, mayweed (*Matricaria perforata*) and chickweed (*Stellaria media*), must be controlled by spraying herbicide every spring if cereal yields are not to be severely reduced.

The origin of pest problems

There are three possible reasons why agricultural ecosystems are so susceptible to pest outbreaks.

1. Agricultural ecosystems are greatly simplified compared to natural ecosystems. At the extreme, monocultures contain only one desired plant species. In such simple ecosystems, organisms can increase in number and achieve pest status whereas their populations could have been controlled by, for example, predators in more diverse ecosystems.

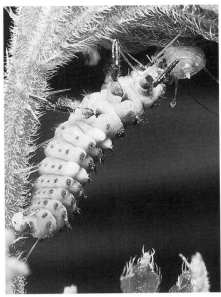

(a)   (b)

Fig 10.3 **(a)** A cottony cushion scale, *Icerya purchasi*, on an orange tree. The tiny, sap sucking insect is actually living under the white, cottony looking material which it has secreted to protect itself.
(b) A ladybird larva eating an aphid. Such larvae are voracious eaters and so, in large quantities, can have a dramatic effect on the number of aphids.

Continually growing the same crop on the same piece of land year after year may also allow pest populations to build up.

2. The unintentional or intentional introduction of a species which, removed from its natural enemies, increases rapidly and reaches unnaturally high densities. For example, cottony cushion scale *(Icerya purchasi)*, a small insect which sucks sap from the leaf and twigs of citrus trees like oranges (Fig 10.3(a)), was introduced into California from Australia sometime in the early 1870s. It caused such damage that by 1887 the entire Californian citrus industry was threatened. The scale insect was eventually controlled by introducing one of its natural enemies *Rodalia cardinalis*, a ladybird from Australia. Encouraging insects like ladybirds (Fig 10.3(b)) to act as biological control agents makes economic as well as environmental sense.

3. Changes in agricultural practice, which may be part of a planned rotation of crops or because of economic changes which suddenly make one crop more profitable than another. Such a switch in crops can cause serious pest problems because of a 'carry over' effect from one crop to the next. For example, the wire worm, which is the larval stage of the click beetle (*Agriotes lineatus*), normally lives in grassland where, although it feeds on grass roots, does no serious economic damage, i.e. it is not a pest. However, if the grass is ploughed and brought into cultivation and used to grow, say, potatoes, then the wire worm will attack the tubers, causing significant economic damage – the farmer now has a pest problem.

QUESTIONS	10.1 Draw up a list of biological properties which you think are likely to make an organism a potential pest.
	10.2 **(a)** Distinguish between the economic injury level and economic damage threshold of a pest.
	(b) Why should farmers not spray their crops as soon as they spot the first pest attacking the plants?

10.3 (a) Why might growing lots of crops together, called intercropping, help to prevent pest problems?

(b) Suggest reasons why intercropping is usually practised in developing countries but not in more developed countries like Britain.

10.2 WEEDS

Plants of both the same and different species can potentially compete for limited resources like light, water, nutrients and space. Such intra- and inter-specific competition reduces survival, growth and/or reproduction. The obvious example of inter-specific competitors in agricultural systems are **weeds**. A weed can simply be defined as a plant growing where it is not wanted. This means that other crops (**volunteers**) can also be weeds. A volunteer is a plant which develops from seeds, tubers or other organs of vegetative reproduction left over from a previous crop.

In addition to the effects of inter-specific competition, weeds cause losses and inconvenience to farmers in other ways (Fig 10.4).

- Some weeds provide alternative hosts for pests and diseases (Table 10.1).

- Weeds can produce compounds which inhibit the germination and growth of crop plants – a phenomenon called **allelopathy**. For example, the leaves of the weed false flax (*Camelina alyssum*) contain phenols which, when they get into the soil, inhibit the growth of the crop plant flax (*Linum usitatissimum*).

- Weed seeds can contaminate the crop after harvesting. This not only reduces its economic value but also represents another source of weeds when the seed is sown to raise a new crop. For example, the seeds of wild

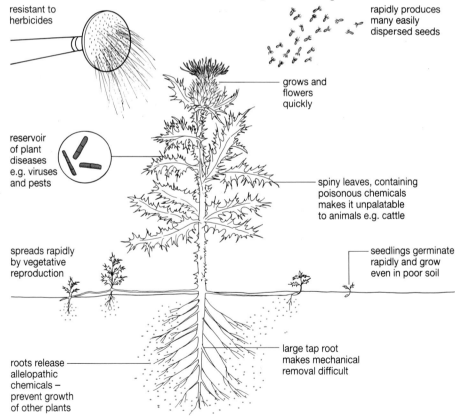

Fig 10.4 The ultimate weed grows and spreads rapidly, produces large quantities of seed, is unpalatable or toxic to animals and inhibits the growth (allelopathy) of other plants. In addition, it is a reservoir for plant diseases and is resistant to herbicides, while its long tap root prevents mechanical removal.

Table 10.1 Examples of weeds acting as hosts for diseases and pests of crop plants.

Type of pathogen or pest	Weed	Crop plant
fungi		
take-all (*Ophiobolus graminis*)	couch grass	cereals
club root (*Plasmodiophora brassicae*)	various members of the Cruciferae	brassicas
viruses		
cucumber mosaic	common chickweed	many crop plants
insects		
black bean aphid (*Aphis fabae* Scop.)	fat-hen, many legumes	broad and field beans

oats (*Avena fatua*) and cleavers (*Galium aparine*) were found in 2 and 8% of wheat samples respectively.

- Weeds may be poisonous or inedible to animals, for example ragworts (*Senecio* spp.) and St John's wort (*Hypericum* spp).

The effect of weeds on crops is shown in Table 10.2. The cost of such competition can be enormous. In countries like Britain such costs manifest themselves as the large bills that farmers have to pay for herbicides. In less developed countries, where the farmers cannot afford herbicides, losses to weeds are still considerable. For example, in Africa, losses to fruit crops due to weeds are estimated at 25%. By comparison the losses to animal pests and disease combined are only 20%.

Table 10.2 Selected figures showing annual losses of crops to animal pests, diseases and weeds. Figures are calculated on a world basis unless otherwise stated, and in the first three lines are in millions of tons.

Crop	Potential production in millions of tons	Losses due to Animal pests	Diseases	Weeds
all cereals	1467.5	203.7	135.3	167.4
sugar beet and sugar cane	1330.4	228.4	232.3	175.1
vegetable crops	279.0	23.4	31.1	23.7
all crops (% of potential value)				
(a) worldwide	100	13.8	11.6	9.5
(b) Europe	100	5.1	13.1	6.8
(c) North and Central America	100	9.4	11.3	8.0
(d) Africa	100	13.0	12.9	15.7
(e) Asia	100	20.7	11.3	11.3

ANALYSIS

Table 10.3

Density (numbers m^{-2})	Yield per plant (g)
10	23.8
150	1.8
450	0.7

Intraspecific competition

This exercise involves using data-analysis and thinking skills.

Look at the data in Table 10.3.

(a) Which variables would you need to control in the experiments needed to obtain these results?

(b) Account for the reduction in yield per plant.

(c) Calculate total yield in gm^{-2} for each of the densities. Comment on the implications of your results for a farmer working out which density to sow expensive seed at to obtain the maximum yield.

Types of weed

There are two basic sorts of weed – annuals and perennials. Annuals complete their life cycle in less than one year; perennial species can grow and reproduce for many years. The two sorts present rather different problems for farmers.

Annuals are characterised by:

- short life cycles and rapid growth rates;
- high seed output (Table 10.4) under ideal conditions and the ability to produce some seed even under poor conditions, e.g. in nutrient-poor soil;

Table 10.4 Average output of seeds of common weeds.

Weed	Average output per plant
groundsel (*Senecio vulgaris*)	1 000 – 1 200
common chickweed (*Stellaria media*)	2 200 – 2 700
greater plaintain (*Plantago major*)*	13 000 – 15 000
common poppy (*Papaver rhoeas*)	14 000 – 19 500
hard rush (*Juncus inflexus*)*	200 000 –234 000

* Perennial weed.

- flowering and seed production after only a short period of vegetative growth;
- long seed longevity, which means that the seeds can remain viable even after long periods buried in the soil. For example, 47% of shepherds purse seeds germinated after 16 years with 84% of greater plantain seeds germinating after 21 years. This means that the seeds can accumulate in the soil to form a **seed bank** (Table 10.5). This represents an agricultural time bomb since these seeds can germinate at any time, for example, when brought close to the surface by ploughing.

Table 10.5 Populations of viable weed seeds of some arable soils in Britain.

Source	History of land	No. of seeds (m^{-2})
Rothamsted	continuous wheat	34 100
Woburn	continuous barley	29 900
Midland clay	farm crops	4000–27 400

- no special germination requirement. Some seeds require special conditions before they will germinate but this is not true of many weeds (Fig 10.5).
- well developed dispersal mechanisms.

All these adaptations mean that annual weeds are well suited to colonise and take advantage of temporarily open land, e.g. a recently ploughed field.

Two types of annual weed can be distinguished: **Winter** and **summer annuals**. Winter annuals have seed which germinates in the autumn. The plants then either flower rapidly, producing seed before the winter, or overwinter in a vegetative state flowering in the following spring. The seeds of summer annuals germinate in the spring, flower in the summer, produce seed and then die.

Perennial weeds (Table 10.6) reproduce both sexually and vegetatively by means of perennating organs, e.g. rhizomes (couch grass) and tap roots (docks). These perennating organs represent an additional problem for the

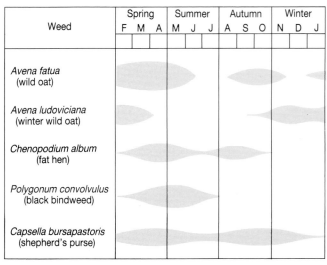

Weed	Spring			Summer			Autumn			Winter		
	F	M	A	M	J	J	A	S	O	N	D	J

Fig 10.5 Germination patterns of annual weeds. Notice how some species only germinate at certain times of year, but others germinate at any time.

Table 10.6 Some characteristics of a few important perennial weeds.

Species	Reproducing parts: over-wintering state	Average depth of vegetatively reproductive parts	Seed production
Aegopodium podagraria (ground-elder)	rhizomes; dormant buds underground overwinter	shallow (0–25 cm)	unimportant
Agropyron repens (common couch)	rhizomes; dormant buds underground but aerial shoots overwinter	shallow	fairly important
Convolvulus arvenis (field bindweed)	creeping roots; these overwinter	very deep (down to >3 m)	very important; only set in warm summer
Ranunculus repens (creeping buttercup)	procumbent creeping stems; some leaves overwinter	above ground	very important
Rumex crispus and *R. obtusifolius* (curled and broad-leaved dock)	tap roots; rosette of leaves overwinters	very shallow	very important

farmer. For example, couch grass, one of the most troublesome of all agricultural weeds in Britain, has rhizomes which contain considerable food reserves. These:

- support vigorous growth at the start of the growing season which gives the plant a competitive advantage;
- sustain the plant through unfavourable climatic conditions;
- enable the plant to survive intense competition with other plants.

In addition, the rhizomes carry many lateral buds which can develop into new plants by vegetative growth. Cultivating soil full of couch grass rhizomes by ploughing or harrowing will split the rhizomes up, spread them all over the field and so increase the weed problem.

PESTS – BIOTIC FACTORS AFFECTING CROP PRODUCTIVITY

10.4 (a) What makes a plant a weed?

(b) Distinguish between and give examples of (i) annual and perennial weeds (ii) summer and winter annual weeds. (**Hint:** use Fig 10.5)

10.5 (a) Explain the biological basis of the saying 'one years seeding ten years weeding'.

10.6 Why is it important for farmers to ensure that they clear their land of old crops before planting new ones?

10.3 ANIMAL PESTS

(a)

(b)

Fig 10.6 **(a)** Peach potato aphids, *Myzus persicae*, attacking a cabbage. The insects suck sap from the plant through fine, pointed stylets which they insert into the leaf veins. Direct damage is usually leaf curl on young leaves, with aphids clustering underneath the leaves. The aphids also transmit viral diseases. **(b)** This cabbage shows the dramatic effect of an infestation of cabbage white caterpillars. All the leaves have been eaten, reducing the plant to the bare branches.

Animals, including insects, molluscs (slugs and snails), birds and rodents (mice and rats), have plagued humans for a long time as the Bible indicates in the Book of Exodus:

... and when it was morning the east wind brought the locusts ... For they covered the face of the whole land, so that the land was darkened and they ate all the plants in the land and all the fruit of the trees ... not a green thing remained, neither tree nor plant of the field, through all the land of Egypt ...

Worldwide, animal pests cause losses, both in the field and in storage, of between 30 and 50% of crop yields. Some crops, for example cotton, are particularly prone to pest attack. Thus, it is estimated that 50% of Pakistan's cotton crop would probably be lost to insect attack each year in the absence of spraying with insecticide.

Arthropod pests

There are so many different sorts of animal pest that we will concentrate on just one major group, the arthropods. In particular, we will look at insects, which are major causes of crop losses in the tropics and subtropics but are less important in temperate regions like Britain (see Table 10.2). Nonetheless, insects, in common with other pests like nematodes, can cause local problems to British farmers under appropriate environmental conditions. For example, the hot, dry summers of 1975 and 1976 led to heavy infestations of cereals and other crops by aphids while recent mild winters have led to increased damage by eel worm (a nematode) to cereals. Two examples of insect pests are shown in Fig 10.6. Note in particular the way in which they attack and damage the plant.

Insects can attack crops directly by damaging the plant part which is to be harvested, for example by feeding on potato tubers or sugar beet roots, or indirectly by attacking a plant part which is not harvested, for example the roots of cereals or the leaves of sugar beet or potato. **Direct damage** by insects with chewing mouthparts includes the following.

- Reducing the amount of leaf tissue, which hinders plant growth by reducing the total amount of photosynthate which can be made (i.e. reduces LAI), e.g. locusts and caterpillars of butterflies and cabbage white.

- Tunnelling into stems, which interrupts the flow of sap in phloem and xylem, often destroying the apical parts of the plants. Such stem borers include wheat stem sawfly (*Cephus pygmaeus*) and the corn borers (*Ostrinia* spp.) which tunnel in the stems of maize.

- Destruction of buds and growing points, for example fruit bud weevils (*Anthomomus* spp.) on apples and pears.

- Causing premature fruit fall, for example codling moth on apples.

- Attacking roots, so interfering with water and nutrient absorption, for example wire worms, leather jackets.

Insects with piercing and sucking mouthparts, like aphids, can cause direct damage in the following ways.

- Reduce plant vigour due to the removal of excessive quantities of sap. In extreme cases, wilting and distortion of the foliage (leaf curl) may result, as caused by aphids on plants like potatoes and peaches.
- Damage floral organs and reduce seed production, for example capsid bugs.

Indirect damage effects include the following:

- Making the crop more difficult to cultivate or harvest through, for example, promoting plant distortion, particularly the development of spreading growth, which makes spraying and weeding more difficult. For example, larvae of the fruit fly (*Oscinella frit*) bore into the stem of the young stems of cereal plants producing a 'dead heart'. This causes the stem to die and turn brown and wither. Profuse tillering then results, producing cereals with spreading growth.
- Reducing the crop quality either through reduction in the nutritional quality of the crop (e.g. pests of stored grain like grain weevils) or marketability. Carrots damaged by the larvae of carrot root fly are practically worthless, with tinned vegetable manufacturers rejecting a whole 10 tonne load if it contains just a few damaged carrots.
- Transmission of disease organisms. This is the most serious indirect effect. In particular, most plant viruses depend on an insect vector, for example an aphid, to transmit them between host plants.

| QUESTIONS | **10.7** Compare and contrast the damage done to plants by insects with chewing and piercing mouthparts. |
| | **10.8** During the Second World War, potatoes were grown continuously on peatland areas, called mosses, in southwest Lancashire. As a result, a high number of cyst nematodes, which attack the tubers, developed.
 (a) Account for the build-up of the nematode numbers.
 (b) Suggest why crop rotation might prove a useful method for controlling this animal pest. |

10.4 PLANT DISEASES

Plant diseases affect the plant's metabolism, for example upsetting the plant's photosynthetic machinery, so the plant cannot grow as fast. While this may not kill the plant the disease may weaken it and so render the plant more susceptible to attack by other pests and diseases. In addition, plant diseases affect the appearance of the plant, so reducing the commercial value of the crop or even making it totally unfit for human consumption (Fig 10.7).

Crop diseases may be caused by fungi, bacteria and viruses (Table 10.7). Such diseases are far more important than animal pests in reducing crop yields in temperate agricultural systems. One estimate from 1973 puts losses from leaf diseases alone at £70 million, while the last estimate of global losses were put at 10–15% of potential crop yield (see Table 10.2). However, in advanced agricultural countries, like those of the European Community and America, crop losses to disease are usually much less than 10% because of the sophisticated control measures available.

Fungal diseases are probably the most important cause of crop losses in temperate regions. Such diseases can be spread either through the soil or through the air (see Table 10.7). **Soil-borne diseases**, like **eyespot**

Fig 10.7 Powdery mildew, *Erysiphe graminis*, growing on wheat ears. Notice how the disease has infected practically all the grains, so reducing the commercial value of the crop. Grain suffering this damage cannot be used to make bread flour because it would have a low protein content and unacceptable odours.

Table 10.7 Some plant diseases and their causative agents.

Disease	Plants affected	Causal organism
barley yellow dwarf virus disease	barley, wheat	virus
beet yellow virus disease	sugar beet	virus
potato blackleg	potato	bacterium, *Erwinia carotovora* subsp. *atroseptica*
potato blight	potato	fungus (soil borne) *Phytopthora infestans*
take all	wheat, barley, oats	Fungus (soil borne) *Gaeumannomyces graminis*
eye spot	wheat, barley, oats	fungus (soil borne) *Pseudocercosporella herpotrichoides*
powdery mildew	wheat, barley, oats	fungus (air borne) *Erisyphe graminis*

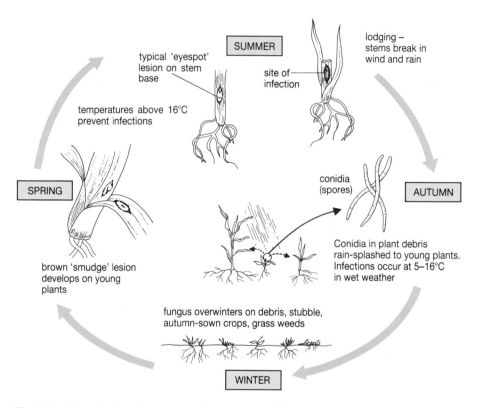

Fig 10.8 Life cycle of a soil bore fungus, *Pseudocerosporella herpotrichoides*, eyespot of cereals. The fungus can be controlled by several methods. For example, disposal of infected stubble and a two year break from growing cereals. Fungicides are also used. New varieties of cereals have been produced which are resistant to eyespot, e.g. the winter wheat variety Rendezvous.

(Fig 10.8) and **take-all** of cereals, usually overwinter on stubble from previous crops or on weeds. Alternatively, the fungus can overwinter on infected seeds, as in the case of **stem canker** of oil seed rape. **Air-borne diseases** are transmitted by spores being blown on the wind or carried by insects. Examples include **powdery mildew** (Fig 10.9) and **stem root** of oil seed rape. The importance of these fungal diseases cannot be over-emphasised. For example, powdery mildew alone is responsible for 34.5% of losses in yield in barley and 20% in wheat in Britain.

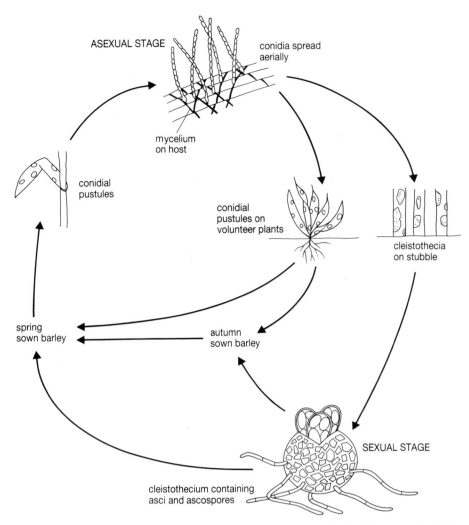

Fig 10.9 Life cycle of *Eryspihe graminis* f. sp. *hordei* (powdery mildew of barley). The asexual conidial spores are produced in vast numbers in warm, dry weather.

Fig 10.10 A leaf infected with sugar beet yellows virus (top) compared with a normal healthy leaf below. Infected leaves are usually small, show symptoms of chlorosis (lack of chlorophyll) and crack easily. The only way to control the disease is to control the aphids which spread it, for example by spraying.

Viral diseases include **potato leaf roll**, **sugar beet yellows** (Fig 10.10) and **barley yellow dwarf** virus. The majority of viral diseases are transmitted by insects like aphids which pierce the plants phloem to feed on their contents. For example, sugar beet yellows is transmitted by aphids, particularly the peach–potato aphid, *Myzus persicae* (see Fig 10.6(a)). After the virus has been 'injected' as the aphid feeds it is translocated throughout the whole plant. The initial symptoms develop as yellowing of the leaf margins followed by yellowing of the whole leaf, resulting in a reduction in photosynthetic efficiency. The virus overwinters in stores (clamps) of mangels and beets from where it is picked up by aphids in the spring. The incidence of the disease varies from year to year. In cold winters, fewer aphids survive and, as a result, the incidence of sugar beet yellows is less than after a mild winter. Losses from yellows are especially severe when sugar beet is planted early. In 1974, a year of severe infection in England, sugar beet yields were 2 million tonnes less than in 1970–3, a reduction of 25%.

Potatoes are particularly prone to viral diseases. In the case of both potato leaf roll and potato virus Y there is a reduction in both yield and crop quality. When the virus is introduced into a plant it rapidly becomes systemic, being carried to all the plant tissues by the plant's own transport systems. In the case of potato leaf roll this leads to an inrolling of the leaves with infected leaflets becoming thick and dry with the lower leaves turning brown. The infected tubers lack vigour and are undersized. A particular

PESTS – BIOTIC FACTORS AFFECTING CROP PRODUCTIVITY

problem with potatoes is that plants are grown from seed potatoes. If these tubers are infected then the plant which grows from them will also be infected. For this reason, seed potatoes are raised in colder areas, like Scotland, where aphids are rarer.

QUESTIONS	
	10.9 How are plant pathogens spread between plants?
	10.10 Why will minimal cultivation methods potentially lead to an increase in the numbers of soil-borne diseases like take-all?

SUMMARY ASSIGNMENT	
	1. **(a)** Make sure you have definitions and examples of summer and winter annual weeds, and perennial weeds.
	(b) How do weeds affect crop productivity?
	2. Produce a table to show how insect pests can be classified according to the type of damage they cause. Include two examples of animals which fall into the categories you identify.
	3. Outline the life history and effects of (i) a viral disease (ii) a soil-borne and (iii) an air-borne fungal disease of plants.
	4. **(a)** Discuss the following proposition: The term pest is an economic one since it has no biological meaning.
	(b) Compare and contrast the biological properties of weeds, insect pests and plant pathogens.

Chapter 11

PEST CONTROL

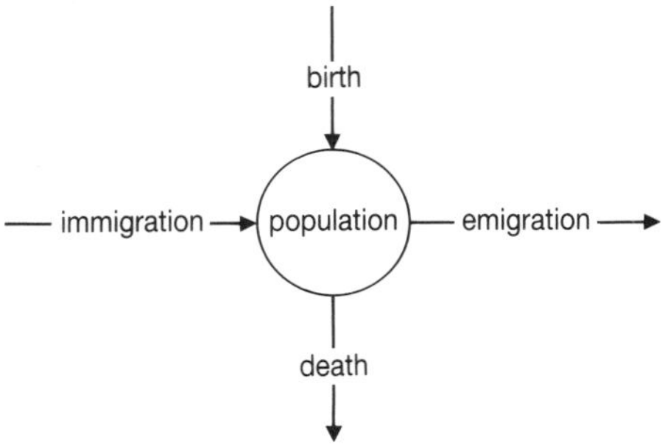

Fig 11.1 The ecological facts of life. Only four processes, birth, death, immigration and emigration, can affect the size of a population.

The aim of all methods of pest control is to keep the size of the pest population below the economic injury level (see Fig 10.2). Since only four processes control the size of any population (Fig 11.1) this can be achieved by:

- increasing the death rate, e.g. by spraying with a pesticide;
- decreasing the birth rate, e.g. by releasing sterile males to control screw worms, a fly which is a serious pest of cattle;
- increasing emigration of the pest, e.g. by using trap crops (see Section 11.3);
- decreasing immigration, e.g. by removing habitats where pests can over-winter and from which they can re-invade the crop.

You should also remember that all methods of pest control have benefit–cost ratios and that farmers will choose those techniques which provide the greatest benefits for the smallest costs.

LEARNING OBJECTIVES

After completing the work in this chapter you will be able to:

1. outline cultural, biological and chemical methods of pest control;

2. discuss the relative merits of different methods of pest control;

3. describe the implementation of a programme of integrated pest management;

4. synthesise information to answer general and specific questions about pest control;

5. appreciate the nature of the decisions that farmers have to make during the course of a growing season.

11.1 PROTECTING CROPS – CULTURAL METHODS

These include crop rotation, weeding and tillage, which are still the most important and widely practised methods of crop protection.

Tillage

One of the main purposes of ploughing and secondary tillage operations like harrowing is to control weeds (see Section 9.1). Such processes uproot weeds and incorporate them into the soil where they then rot down. Discing and harrowing destroy seedlings and reduce vegetative growth of perennial rhizomes. However, several such operations will be necessary to control perennial weeds like couch grass and docks. On a smaller scale, hand weeding and hoeing may be used. Tillage is also important in controlling animal pests and diseases (Fig 11.2). Thus tillage brings soil pests like wire worms to the surface, exposing them to predators, while the incorporation of crop residues and weeds into the soil removes habitats where pests and diseases can overwinter.

Fig 11.2 Ploughing helps to control all forms of pests. The seagulls will be eating soil pests such as the larvae and pupae of insects, which are over wintering in the ground. Note how weeds and crop residues are buried. How will this help to control fungal diseases?

Crop rotation

Continuous cultivation of crops on the same piece of land allows the build-up of diseases in the soil, for example take-all and eyespot, important fungal diseases of cereals. Such soil diseases are best controlled by crop rotation, where continuous cropping of cereals, for example, is interrupted by planting grass or root crops like potatoes, which break the disease cycle. Crop rotation is also successful in controlling weeds like wild oats which are not easily eradicated by herbicides. This is best achieved by putting the land down to grass. Similarly, soil pests, like nematodes which attack potatoes, can be at least partially controlled by using crop rotation. Note that the aim is not to eradicate the pest but rather to reduce it to a level where it will do less economic damage. However, crop rotation will only work when a pest cannot colonise successive crops. For example, a common type of rotation is to alternate cereals with a legume. This is effective against some pests but not others, for example the grape colapsis, a beetle, can feed on both types of crop. In Europe a combination of potato, wheat and oil seed rape is a popular and successful rotation for controlling pests.

Fertilising, liming and draining

Improving the soil by fertilising, liming and draining may also help to control or reduce the effect of pests. Thus plants grown in well maintained soil will grow more vigorously and so will be better able to withstand the attack of pests. Liming, by increasing soil pH, may also help to protect plants, for example by controlling club root, a fungus which attacks the roots of brassicas. Drainage helps to prevent the build-up of organisms like slugs which are favoured by wet conditions.

Time of sowing and harvesting

By sowing early or late it may be possible to avoid the egg-laying period of a pest, or the vulnerable stage in plant growth may have passed by the time the pest reaches pest proportions. For example, autumn sowing of cereals in Europe has reduced the risk of aphid damage to seedlings to a minimum while the dense growth of an autumn sown crop also reduces weed problems in the following spring. However, autumn sowing does increase the risk from wheat bulb fly (*Delia coarctata*). Late sowing of spring wheat, on the other hand, reduces the risk from bulb fly.

Early harvesting may prevent crops from being infested with a pest and early maturing varieties are one of the major goals of plant breeders. As

Fig 11.3 A barrier of fine netting has been placed around this carrot crop to prevent attack by carrot root fly. The barrier only needs to be about 0.5 m high as the fly flies near the ground in her search for carrots in which to lay her eggs. It is the larvae which damage the carrots.

with early/late sowing the idea is to desynchronise the growth of the crop from the life cycle of the pest. Thus infestation by maize weevil (*Sitophilus zeamais*) is reduced by prompt harvesting of the maize crop before the adult beetles emerge from stored maize.

Other methods

Crop sanitation involves:

- the destruction of diseased or badly damaged plants;
- removal and destruction of plant rubbish, e.g. fallen fruits;
- the destruction of crop residues.

It is essential for pest and disease control. For example, the European corn borer overwinters in the stubble of maize while the rotting residues of turnips, parsnips and brussel sprouts attract bean seed flies.

Physical barriers (Fig 11.3) used also be effective against some pests. For example, codling moth used to be successfully controlled in apple orchards by the use of sticky bands around the trunk of the tree.

Trap crops can be used to lure pests off crop plants. This involves planting plants which the pest prefers to the crop plant. The pest can then either remain and feed on the trap plant or the trap plant can be sprayed with insecticide. Since the trap plants are only planted as a peripheral band, or at about every fifth to tenth row, this represents a major saving in insecticide use.

Intercropping (Fig 7.16) is a variation on this theme where two crops are planted together. For example, growing strongly scented plants like onions around carrots helps to protect the latter from carrot root fly, which is attracted by the scent of carrots. Alternatively, the pest can be lured off the main crop plant and then destroyed. In northeast Thailand in 1980 ground nuts were intercropped with maize with the effect that nymphs of the Bombay locust (*Patanga succinata*) left maize, particularly on hot days, to find shade on the lower leaves of the ground nut plants. Here they were eaten by ducks.

Mulching involves covering the soil with organic material like animal manure, or black polythene sheet. This inhibits weed growth by reducing the light available for photosynthesis.

Stubble burning destroys both weeds and their seeds, fungal spores and animal pests but is now strictly controlled.

The importance of cultural methods should therefore not be under-estimated. However, cultural methods can be labour intensive and may involve growing crops as part of a rotation which are not economically valuable. In intensive agricultural systems the use of chemical methods of pest control have therefore gained in popularity.

QUESTIONS

11.1 (a) How might the development of systems of minimal cultivation result in an increase in pest problems to the farmer?
(b) Suggest how these problems might be overcome.

11.2 (a) How might farmers 'desynchronise' the growth of crops from the life cycle of the pest attacking them?
(b) What are the advantages and disadvantages of such methods of pest control?

PEST CONTROL

11.2 CHEMICAL CONTROL – HERBICIDES AND FUNGICIDES

Modern agricultural research has supplied the farmer with a battery of toxic chemicals which can be used to control pest organisms. These include:

- herbicides – for the control of weeds;
- fungicides – for the control of fungal diseases;
- insecticides – for the control of insect pests;
- nematicides – for the control of nematodes.

Herbicides

Herbicides (Table 11.1) are not a new invention. In the Second World War some weeds in cereal crops were controlled by spraying the fields with sulphuric acid solution prior to planting. However, a second agricultural revolution was sparked in the 1940s with the discovery of the selective herbicidal activity of the phenoxyacetic acids, 2, 4–D and MCPA. These were toxic to broad-leaved weeds, like the corn poppy (*Papaver rhoeas*) and charlock (*Sinapsis arvensis*) which were then important weeds of cereal crops, but did not damage the cereal plants. Today 75% of all chemical pesticides applied to crops in Britain are herbicides.

Table 11.1 Commonly used herbicides.

Type of herbicide	Examples	Mode of action	Main uses
phenoxy acids	MCPA; 2,4–D	growth regulator	selective herbicide in cereals
carbamates	Propham	growth inhibitor	pre-emergence control in sugar beet
heterocyclic nitrogens	simazine, atrazine	photosynthesis inhibitor	total control of herbaceous weeds (high concentration), selective control of annual weeds in maize, field beans
phenylureas	Linuron, diuron	photosynthesis inhibitor	pre- and post-emergence control of annual weeds
trichloroacetic acids	TCA	growth inhibitor	control of couch grass and wild oats
nitrates	Ioxynil	growth inhibitor	post-emergence control of cleavers, mayweeds in cereals
bipyridilums	paraquat, diquat	photosynthesis inhibitor	general weed control

Pre-emergence herbicides are applied to the soil prior to the emergence of the crop. Examples include:

- **Contact herbicides** like paraquat and diquat which kill all above-ground parts of any plant. These are particularly important in systems of minimal cultivation, e.g. direct drilling, since their rapid action and rapid breakdown in the soil mean that the crop plant can germinate into a weed-free and herbicide-free environment.

- **Residual herbicides** which are applied to the soil. They bind to the soil particles selectively killing weed seedlings as they emerge. An example is

Linuron which binds to the thylakoid membranes of chloroplasts and inhibits photosynthesis.

Post-emergence herbicides are applied to the leaves of both the crop and the weed. Consequently, they need to be selective in their action. Examples include **growth regulators** like 2,4–D which are absorbed by the leaves of dicotyledenous weeds, translocated throughout the plant and promote abnormal growth. 2,4–D, for example, inhibits cell division and elongation, thereby preventing root extension. Roots lose some of their ability to take up nutrients and water, photosynthesis is reduced and the efficiency of transport in the phloem is inhibited.

Systemic herbicides can be applied to either the soil or the plant's leaves. They are absorbed and translocated to the growing points (meristems) in the plant. Examples include glyphosate, a **non-selective** systemic herbicide, which is applied to fields after harvesting and prior to sowing, killing both annual and perennial weeds. Asulam is a selective systemic herbicide which acts by inhibiting cell division. It is used for controlling docks (*Rumex* spp.) in grassland and bracken.

Herbicides are applied as sprays with water as the carrier (Fig 11.4). The problem with all spraying techniques is that of **drift** – the spray is blown by the wind into areas where it is not needed. For example, paraquat sprayed on to potato fields in the spring may blow on to winter wheat, causing the wheat plants to be scorched and perhaps killing them. In addition, the spray can get into rivers, into woodland areas or into housing estates with potentially catastrophic side-effects.

Fig 11.4 Herbicides being sprayed onto spring barley for control of broad leaved, dicotyledenous weeds. Notice how the spray is drifting.

Fungicides

The farmer has a wide variety of fungicides available (Table 11.2), which can be applied to seeds (seed dressings) or sprayed on to crops once they have germinated. Cereal seeds, for example, may be treated with organic mercury compounds or carboxin to prevent bunts and smuts. Alternatively, fungicides can be sprayed directly on to the leaves where most, for example draxolon, act on contact with the fungus, i.e. they are non-systemic or **protectant**. Others, like benomyl and ethirimol, are systemic, entering the plants tissues after absorption from the leaves. The effectiveness of fungicidal application depends on a number of factors, including the rate of application, the activity of the compound, the persistence of the compound and, more importantly, the timing of the application. Sprays against fungi are normally only applied after the farmer has received a specific warning from, for example, ADAS. Usually, such a warning is based on prevailing and predicted weather conditions. For example, the development of powdery mildew is favoured by dry, warm conditions; the advent of such weather conditions will trigger a warning and the farmers will spray.

Table 11.2 Some commonly used fungicides.

Fungicide	Main uses	Application and action
Ethirimol	cereal mildews	seed dressing, systemic
Benomyl	eyespot	spray, systemic
Organomercury	bunts, smuts	seed dressing, protectant
Carboxin	cereal smuts	seed dressing, systemic
Thiram	damping off	seed dressing, protectant

The benefits and costs of using fungicides

This is a data-analysis exercise.

Table 11.3   Yield response of winter wheat to fungicide application in Herefordshire.

Variety of wheat	1980	1981	1982	1983
A	1.67	1.81	1.61	2.76
B	1.21	1.95	2.14	2.10
C	–	1.81	2.05	1.78
D	1.97	2.30	2.41	2.62
E	2.73	1.89	2.77	3.43
F	–	2.44	3.21	2.67

Table 11.4   Mean increase in yield in 1979–80 from using various applications of fungicide.

Application of fungicide	Yield increase (t ha^{-1})
1 (early season)	0.52
1 (middle season)	0.83
1 (late season)	0.63
2 (early and middle)	1.07
2 (early and late)	1.08
3 (early, middle and late)	1.63
4 (early, middle, late and pre-harvest)	1.58

Yield without fungicide = 6.38 t ha^{-1}.

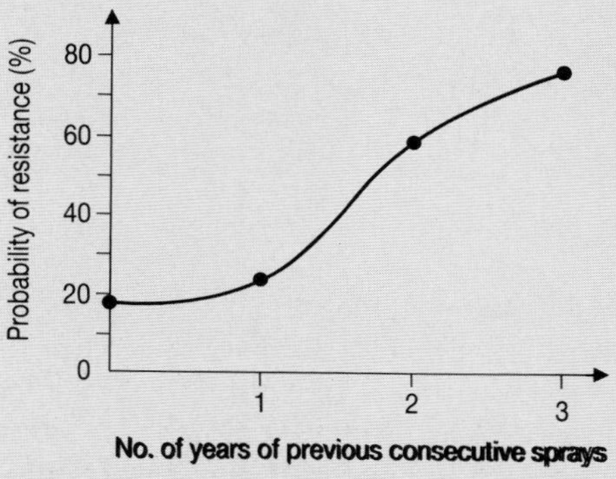

Tables 11.3 and 11.4 give some data on the response of wheat to the application of fungicide.

(a) Is the increase in yield which results from fungicide treatment the same from year to year? Suggest reasons for your answer.

(b) Do different varieties of wheat respond equally well to fungicide application? Explain your answer.

(c) What problem does Fig 11.5 suggest is associated with continual use of fungicides? Do you think this will be a problem with all pesticides? Explain your answer.

Fig 11.5   The effect of previous use of fungicide sprays on the development of resistance in eyespot (*Cercosporella herpetrichoides*)

Problems with the use of herbicides and fungicides

While it is the use of insecticides which causes most concern in terms of their effect on non-target organisms, herbicides and fungicides contain potentially toxic chemicals which can get into food chains. For example, the herbicide 2,4–D is now known to be a carcinogen while 2,4,5–T is a known teratogen (causes birth defects). Furthermore, organomercury compounds used as fungicidal seed dressings can accumulate in food chains to toxic levels and anyone eating the seed by mistake will be poisoned (see page 176).

Again, while not as great a problem as with insecticides, weeds and fungi do become resistant to herbicides and fungicides. For example, at least seven groundsel (*Senecio vulgaris*) populations in Great Britain are resistant to triazine herbicides like Simazine. Some examples of fungicidal resistance are given in Table 11.5. Strategies for overcoming the problems of resistance are given in Section 11.3. The genetics of the process are discussed in another book in this series, *Applied Genetics*, by Geoff Hayward.

Table 11.5 Examples of resistance to systemic fungicides.

Fungicide or group	Date of introduction for control of pathogen specified	Pathogen initially controlled	Resistance detected in field populations of pathogen
Dimethirimol	1968	*Sphaerotheca fuliginea* (cucumber powdery mildew)	1970
Ethirimol	1969	*Erysiphe graminis* f.sp. *hordei* (powdery mildew of barley)	early 1970s
Triazole EBIs	1975–6	*Erysiphe graminis* f.sp *hordei*	early/mid-1980s
Acylalanines	1978–9	*Bremia lactucae* (lettuce downy mildew)	mid-1980s
Acylalanines	1978–9	*Phytopthora infestans* (potato blight)	early 1980s

---

QUESTIONS

11.3 Distinguish between the following pairs of terms:
(a) selective, non-selective herbicide
(b) pre-emergence and post-emergence herbicide
(c) systemic, contact herbicide/fungicide.

11.4 Using the information in this and the previous chapter, list the ways in which weeds can directly and indirectly reduce crop yields.

11.3 INSECTICIDES

These can be classified as either **broad spectrum**, which kill a wide range of insects, often including the pest's natural enemies, or **narrow spectrum** or bio-rational, which only kill specific pests. Broad-spectrum insecticides include inorganics, chlorinated hydrocarbons (organochlorines), some organophosphates and some carbamates. Narrow-spectrum insecticides include some organophosphates and carbamates, microbials (dealt with under biological control), insect growth regulators and semiochemicals.

Mode of action

Insecticides can act in a number of different ways.

- **Fumigants** (volatile compounds like phosphine gas) and **smokes** (insecticidal powders mixed with a combustible material) are used in enclosed spaces like greenhouses and grain stores They enter the insect via its tracheal system.

- **Stomach poisons** have to be ingested to poison the insect. They are usually sprayed on to foliage where they are eaten by leaf-eating insects, e.g. rotenrone (derris) used to control caterpillars.

- **Contact poisons** are usually absorbed directly through the insect's cuticle. They may be short lived (**ephemeral**), breaking down rapidly to harmless by-products, e.g. pyrethrum. These are usually applied as foliar sprays. Alternatively, contact poisons may be **persistent** (long lived), like cypermethrin, a synthetic pyrethroid, or DDT, an organochlorine. They can be applied to both the soil or to the leaves.

- **Systemic poisons** are absorbed and translocated by the plant and are particularly effective against sap suckers. They may be applied as sprays or as granules to either the leaves or soil. Examples include malathion (an organophosphate) and primicarb, a selective carbamate insecticide, which kills aphids but not the ladybirds which feed on them.

Types of insecticide

A bewildering array of insecticides are available. The main types include the following.

Pyrethroids – pyrethrins, extracted from the flowers of *Pyrethrum cinerariaefolium*, are powerful non- systemic, contact and stomach poison insecticides which cause rapid paralysis or 'knockdown' of insects by interfering with nerve impulse transmission. They rapidly decompose in the environment which makes them particularly useful as household insecticides. For agricultural use, synthetic pyrethroids, such as cypermethrin and deltamethrin, which are much more stable and so persist on the foliage, are widely used, for example against leaf-eating and fruit-eating moths and beetles. However, they are very poisonous and dangerous to non-target insects like bees and fish.

Chlorinated hydrocarbons (organochlorines) are broad-spectrum, persistent contact insecticides. They act by binding to the membranes around axons, so inhibiting nerve transmission. Insoluble in water but highly soluble in lipids these compounds tend to accumulate in fatty tissue, leading to problems of bioaccumulation (see Chapter 15) and long-term persistence in the environment. Examples include DDT, toxaphene, aldrin, dieldrin, lindane, methoxychlor and chlordane. Their persistence makes them particularly useful against soil pests, e.g. fly larvae.

Organophosphates were developed during the Second World War partly as a result of research on nerve poisons. They are much more poisonous than organochlorines and act by inhibiting the enzyme acetylcholinesterase which is essential for nerve impulse transmission across synapses. Their advantage over chlorinated hydrocarbons is that they are much less persistent in the environment and their higher solubility in water means that they can be translocated through plants, providing a systemic action. This makes them particularly effective against aphids which feed on phloem sap. Examples are malathion, parathion and diazinon. Organophosphates are extremely toxic, many of them to humans, and therefore need to be handled with great care.

Carbamates work in a similar way to organophosphates but are much less toxic to mammals. Several are translocated in plants, providing a systemic action. Unfortunately, most carbamates are highly toxic to bees, essential for pollination, and parasitic wasps, which are natural predators of the pests. The most widely used carbamate is carbaryl, a broad-spectrum, contact insecticide which is particularly effective against chewing insects like caterpillars. There is sufficient difference in the structure of insect acetylcholinesterases to allow agricultural chemists to produce more specific carbamate insecticides. For example, primicarb is a systemic insecticide absorbed through the plant's roots which kills aphids but not the ladybirds which feed on them.

Insect growth regulators are designed to mimic the action of natural insect hormone, so interfering with the development of the insect. Their advantage is that they are highly specific. In particular, they do not affect vertebrates although they may act against insect predators who would normally prey on the pest. Two examples are:
- diflublenzron, a chitin synthesis inhibitor, which prevents the formation of the insect's exoskeleton when it moults;

- methoprene, a compound with a similar structure to insect juvenile hormone (a so-called analogue), which prevents the insect from developing into an adult. This will reduce the size of the pest population in the next generation.

Semiochemicals are naturally occurring compounds which act by changing the behaviour of the insect rather than killing it directly. They include pheromones, chemicals secreted into the air or on to surfaces which affect the behaviour of the same species of insect. For example, sex attractant pheromones are used commercially to control moth pests by interfering with their mating. Another pheromone which is attracting a lot of attention is the aphid alarm pheromone (E) – β – farnesene. This substance makes aphids move around more on the plant and is used to increase the effectiveness of a fungal pathogen (*Verticillium lecani*) which is used to control aphids infesting greenhouse chrysanthemum.

Application

Insecticides can be applied as:

- **sprays** usually with water as the carrier. The low solubility of insecticides in water has led to the development of special formulations to ensure that the insecticides disperse evenly in the water. For example, wettable powders consisting of the insecticide attached to a carrier particle, usually clay, and a wetting agent. The major problem with sprays is that they drift and so get into areas where they are not needed or desired, e.g. housing estates and rivers. With an insecticide like cypermethrin which is very toxic to fish the results may be disastrous.
- **dusts** (Fig 11.6) which consist of the insecticide mixed with a carrier like clay. Again, drift is a problem.
- **granules** of insecticide and small pellets of clay (20–80 µm in diameter) to the soil. For example, when carrot seed is planted, granulated organophosphate insecticide, e.g. chlorfenvinphos, is added at the same time to control carrot root fly, the larvae of which bore into the developing carrot roots.
- **seed dressings** which can be liquids adsorbed on to the seed coat or powders which stick to the seed coat. They are used against soil pests like wire worms, chafer larvae and shoot flies.
- **baits** which consist of an insecticide like dieldrin or heptachlor mixed with an attractant, such as an edible substance like meal, plus a carrier and preservatives. Baits are used against insects like ants, e.g. Mirex 450, and cockroaches. They are also used against rats and birds. They are mainly used in storage areas but are now being increasingly used in field application.

Fig 11.6 Grape vines being dusted with insecticide. The chemical is attached to the outside of the fine particles which you can see blowing in the turbulence around the helicopter blades.

Problems with insecticides

Insecticides are highly toxic and their over and non-selective use has led to major ecological problems.

Toxicity to non-target species

Insecticides may kill a wide range of organisms rather than just the target organism, particularly when the pesticide persists in the environment, e.g. DDT. Careless application of pesticide may also mean that the chemical spreads far beyond the site requiring treatment. This may occur, for example, if a crop is sprayed on a windy day when the insecticide may drift into adjacent areas. The potential for damage is well illustrated by the application of large doses of dieldrin (an organochlorine insecticide) to

Illinois farm land between 1954 and 1958 in an attempt to eradicate a pest of grassland, the Japanese beetle (*Popillia japonica*). The pesticide effectively eliminated several species of bird, for example meadow larks and robins, killed 90% of the farm cats and caused severe poisoning in cows and sheep. Many insecticides are also very toxic to humans and so great care needs to be taken with their use. Unfortunately, illness and death caused by the abuse of insecticides is still common, particularly in developing countries where statutory guidelines for insecticide use are poorly developed. In addition, illiteracy, which prevents labels and warnings from being read, is widespread.

Target pest resurgence

In 1953 the cabbage aphid (*Brevicoryne brassicae*) was sprayed for the first time in England with the broad-spectrum insecticide para-oxon. Initially, the results appeared spectacular: the pest was almost completely eradicated. However, within two weeks there was the largest outbreak of cabbage aphid ever seen in England. This phenomenon, called **target pest resurgence**, has been seen repeatedly after application of broad-spectrum insecticides. It is due, primarily, to the pesticide killing both the pest and the insects which prey on the pest and its competitors. As a result any pest individuals which survive the initial chemical onslaught, or which subsequently migrate into the treated area, find themselves in an environment where there is a plentiful food supply, no competitors and no predators. Explosive population growth of the target pest is often the outcome.

Secondary pest outbreaks

It is very unlikely that a crop will be affected by only one species of pest. Rather there will be a **pest complex**, consisting of several species, attacking it. Following application of a pesticide designed to eliminate one pest from the pest complex it is quite common for a second species to reach pest proportions. This occurred, for example, in 1950 when cotton in Central America was sprayed with organochlorine and organophosphate insecticides in an attempt to control two insect pests, the boll weevil (*Anthomomus grandis*) and the Alabama leaf worm (*Alabama argillacea*). While these two species were successfully controlled three other species, the cotton boll worm (*Heliothis zea*), the cotton aphid (*Aphis gossyypii*) and the false pink boll worm (*Sacadodes pyralis*) had reached pest proportions by 1955. By 1960 the number of secondary pest species had increased to eight with the number of pesticide applications needed per year to control the pests having increased from 5 to 25.

Resistance

The application of a pesticide represents an enormous selection pressure being applied to the pest population. Within such a pest population there may be one or two individuals which are resistant to pesticide, for reasons of inheritance. These individuals will survive the pesticide application, breed and so pass on their resistance genes to their offspring. Given the rapid rate of reproduction among insects and many weeds, pest populations which are resistant to the effects of pesticides can evolve very rapidly. The genetic details of the process are reviewed in another book in this series, *Applied Genetics*, by Geoff Hayward. By 1987, 50 species of weed, 150 species of plant pathogen (disease- causing organisms) and 500 species of insect were resistant to one or more pesticides. The phenomenon of multiple resistance is particularly worrying. For example, two types of insect, the cotton boll worm and tobacco bud worm, are resistant to a number of the more commonly used insecticides, including DDT, toxaphene, endrin, methyl parathion and synthetic pyrethroids.

Fig 11.7 Pesticides are dangerous chemicals. Here the operator is in full protective clothing, including a face mask, and using an applicator which reduces drift.

Reducing risks from insecticides

The damaging effect of insecticides on non-target organisms, including people, and the problems associated with, for example, pesticide resistance have led to the development of statutory guidelines and alternative strategies of insecticide use. These include:

* setting minimum safety standards for workers handling insecticides, e.g. the use of protective clothing (Fig 11.7).

* the increased use of non-persistent insecticides, e.g. synthetic pyrethroids and organophosphates.

* the increased use of selective insecticides.

* better toxicity testing to establish the effects of insecticides on non-target organisms like bees.

* better monitoring techniques to ensure that pests have reached their economic damage thresholds before spraying starts. For example, monitoring the number of pea moths in East Anglia using pheromone traps has resulted in a 40% reduction in insecticide use against this pest.

* the development of better application methods, e.g. seed dressings and baits, which localise the insecticide so that it only kills the target organism.

* delaying resistance development by (i) reducing insecticide use (ii) using a series of insecticides so that the pest population is never exposed to any one chemical long enough to develop resistance (iii) by using **synergists**, chemicals which inhibit resistance mechanisms or increase the toxicity of the insecticide (iv) spraying different crops which are host to the same pest with different pesticides (see (ii)) (v) careful monitoring of the development of resistance.

Despite their disadvantages, pesticides remain in widespread use for the following reasons.

* They act rapidly, which is essential when keeping a pest below its economic injury level once it has passed its economic damage threshold.

* Manufacturers have stayed one step ahead of the pest by producing new insecticides, herbicides and fungicides.

* Formulation and application methods have improved so pesticides are now better targeted.

* The benefit–cost ratios of applying pesticides still remain favourable. For example, in the United States every $1 spent on pesticides results in $3.12 of increased agricultural production.

* The demand in developed countries for pest-free and undamaged agricultural produce.

* The need to control pests like locusts in developing countries with limited foreign exchange inevitably results in the use of cheaper, more toxic pesticides which are often banned in more developed countries.

QUESTIONS

11.5 **(a)** Distinguish between and give examples of broad-spectrum and narrow-spectrum insecticides.
  (b) Discuss the advantages and disadvantages of using insecticides.
  (c) For controlling a soil pest, like carrot root fly, would it be best to use a **persistent** or an **ephemeral** insecticide. Explain your answer and define the two terms in bold.
  (d) Why would you want an insecticide sprayed on to the leaves of, say, lettuces to be ephemeral? What are the disadvantages of using such short-lived insecticides?

11.6 Suggest how an acetylcholinesterase inhibitor like primicarb can be selective, killing aphids but not ladybirds.

11.7 Aldicarb is a carbamate which is applied directly on to the soil as granules. Once the granules have dissolved the crops are protected against sap-sucking insects like aphids and whitefly for up to 80 days.

(a) What is the advantage in applying aldicarb in the form of insoluble granules?

(b) Explain how this insecticide is able to control aphids and whitefly feeding on the stem and leaves of the treated crop.

(c) Aldicarb also kills soil nematodes. What does this imply about its mode of action? How is it able to control such different types of animal?

(d) Suggest why carbaryl cannot be used on certain crops like cotton.

11.8 Organochlorine insecticides are cheap and effective and are widely used in developing countries.

(a) Do you think this practice is ecologically desirable? Explain your answer.

(b) What social problems might be caused by totally banning the use of such insecticides?

11.4 BIOLOGICAL CONTROL

This involves controlling the pest by:

- adding a natural enemy, e.g. a predator to the agricultural system;
- removing habitats in which the pest can overwinter or survive during periods when the crops it attacks are not present in the system;
- genetic manipulation of the crop or pest. (This is dealt with in another book in this series, *Applied Genetics*, by Geoff Hayward.)

Natural enemies used successfully as biological control agents include the following.

- **Herbivorous insects** to control weeds. For example, the weed *Hypericum perforatum*, which poisoned cattle leading to poor livestock yields, was controlled in California by the introduction of two species of beetle, *Chrysolina hyperici* and *C. quadrigemina*. The weed was reduced to 1% of its original range in ten years, leading to an improvement in livestock production worth $20 million in the seven years from 1953 to 1957.

- **Parasitoids**, insects which lay their eggs in the larvae or eggs of other insects. For example, in Brazil the sugar cane borer, the larva of the moth *Diatrea saccharalis*, is being controlled successfully by four parasitoids, three species of fly and one wasp. In Britain, whitefly, a common greenhouse pest, can be successfully controlled by a parasitoid wasp, *Encarsia formosa* (Fig 11.8). Such parasitoids can be raised in enormous numbers and then released.

- **Predators** can be used to control a wide number of pests. For example, any crop which has not been sprayed with a broad-spectrum herbicide will contain predatory mites, spiders, ground beetles, staphylinid beetles, ladybirds, various bugs and the larvae of hoverflies, lacewings and ladybirds. In addition, exotic predators may also be introduced. For example, the bug *Cytorhinus fulvus* successfully controlled the leaf hopper *Tarophagus prosperina* in Hawaiian fruit plantations by eating the leaf hopper's eggs.

- **Fungi** can be used to attack both weeds and animal pests. For example, the rust fungus (*Puccina chondrillinae*) is used as a biological control agent

Fig 11.8 *Encarsia formosa* laying her eggs.

for skeleton weed (*Chondrill juncea*), a weed imported into Australia from the Mediterranean. Spores of the fungus *Colletotrichum gloeosporiodes* are sold as a commercial preparation called Collego to control the weed North Joint vetch (*Aeschynomere virginica*), a major pest in Arkansas rice fields. The spores are mixed with water and sprayed on to the rice crop, where it kills 99% of the weed. The fungus *Verticillium lecani* is used to control aphids in greenhouses, while the fungus *Dactyella oviparasitica* attacks the eggs of root-knot nematodes to such an extent that it can exert good control of this serious pest of peach trees.

- **Microorganisms** (bacteria, viruses and protozoa) are also attracting attention. For example, the bacterium *Bacillus thuringiensis* is proving to be a useful control agent against caterpillars of moths and has been used successfully to clear thousands of square kilometres in West Africa of the flies (*Simulium* spp.) which carry the disease river blindness. The bacterium is bought as a powder, mixed with water and a surfactant (a chemical which helps the bacterial preparation to stick to leaves) and is then sprayed on to the leaves in the same way as an insecticide. Caterpillars, for example, start to die within a few days of eating the leaves covered with bacteria. The groups of viruses attracting the greatest interest are the nuclear polyhedral viruses (NPV) and granulosis viruses. For example, in Canadian spruce plantations the larvae of the European sawfly (an insect related to wasps) have been controlled using NPV, while the use of viruses to control moth caterpillars which attack coniferous plantations is being actively researched by, among others, the Institute of Virology at Oxford. Here, considerable effort is now being put into genetically engineering NPVs to make them more pathogenic.

The beauty of all these biological control agents is their specificity and the fact that, since they persist in the environment, they will control the pest for long periods of time, hopefully indefinitely. However, you should not get the impression that biological control succeeds every time it is tried. Indeed, 50% of introductions of exotic natural enemies to control a pest have failed. For biological control to succeed careful research of the problem is needed to identify which natural enemy might succeed in controlling the pest. This takes time and is expensive. Furthermore, since the population of a biological control agent may take a long time to build up in the field the pest may exceed the economic injury level before it is brought under control. The specificity of biological control agents is also a problem. An agent which attacks only one pest is of little use where a range of pests are attacking a crop. Their specificity also means that if the pest dies out then the control agent also goes extinct. The pest can then reinvade and reach pest proportions before the biological agent can be re-introduced.

In the short term, then, biological control is not cheap. However, its long-term control of the pest, the increasing cost of pesticides, pesticide resistance and the ecological effects of pesticide misuses are making it an increasingly attractive option. The benefit–cost ratios of biological control agents have been estimated as $1:$30, and a recent report from the United States suggests that farmers there could cut the use of pesticides by half by increasing their use of biological and cultural methods of crop protection.

QUESTIONS	11.9 Compare and contrast the use of broad-spectrum pesticides and biological control agents for controlling pests.
	11.10 **(a)** Why do you think biological control is most successful in enclosed growing spaces, like greenhouses, rather than in open field situations?
	(b) Why would a biological control agent which totally eliminated a pest not be particularly useful?

11.5 INTEGRATED PEST MANAGEMENT

In 1967 the Food and Agricultural Organisation of the United Nations (FAO) defined a form of pest control now called integrated pest management (IPM).

A pest management system that, in the context of the associated environment and the population dynamics of the pest species, utilises all suitable techniques and methods in as compatible a manner as possible and maintains the pest population at levels below those causing economic injury.

The key idea underlying this philosophy of pest control is to maintain pest populations below their economic threshold (see Fig 10.2), not to eradicate the pest. This means working with the pest's natural enemies and only using chemical sprays when the pest exceeds some threshold value. Thus while broad-spectrum insecticides are used in an IPM system they are used very sparingly.

Developing an IPM programme

The following steps, choices and decisions have to be made when developing an IPM programme.

- Which crops to grow? Obviously, physical factors like climate and soil set the overall limits but within these limits the farmer can choose varieties and types of crops which are resistant to the pests likely to be encountered. However, there may be an economic loss – resistant varieties may be more expensive to buy or have lower yields.

- Which pests are likely to be encountered? Which are likely to cause the greatest losses? This requires detailed understanding of the pest's biology, the setting of economic damage thresholds for each pest and monitoring the levels of pests in the field.

- Selecting non-pesticidal methods of control, e.g. physical methods, cultural methods, varietal methods (the use of resistant crop varieties), biological control. How are they best used? Together or singly? Will they keep the pest(s) below its economic damage threshold? What effect will they have on native parasites, predators and pathogens which attack the pest? How will they affect local habitats, wildlife and amenity value?

- Computer modelling is then needed to integrate all the information collected and to answer questions like 'If we adopt strategy X under conditions Y with crop Z what will happen?'

- Implementation and monitoring. Was the IPM programme successful/unsuccessful? How could it be adapted/improved?

- Alter and reassess the IPM programme.

Case study: The San Joaquin Valley IPM programme

This valley in California is a major cotton growing area. Cotton is attacked by a large range of pests but in the San Joaquin valley the main insect pests are:

- the lygus bug (*Lygus hesperus,*) which feeds on the cotton buds, delaying fruit (boll) formation and hence leading to a reduction in yield;

- the cotton boll worm;

- the beet army worm (*Spodoptera exigua*), a moth larva;

- the cabbage looper (*Trichoplusia ni*), another moth larva.

This complex pest system had been treated with insecticides leading to target pest resurgence, secondary pest outbreaks and the evolution of resistance. The IPM programme in this valley was therefore established to reduce the amount of insecticide being sprayed, so preventing the secondary outbreaks. In addition, the programme needed to avoid insecticides like methyl parathion and carbaryl which are known to reduce cotton yield.

The first step was to set an economic damage threshold for the main pest, the lygus bug. Since this insect only causes damage when the cotton plants are budding it is only necessary to control the bug during this season. Initially, the threshold was set at 10 bugs per 50 net sweeps during the budding season. The plants were only sprayed if the lygus bug exceeded this density. Outside the budding season there was no insecticide application to control the lygus bug. In addition, interplanting cotton with strips of alfalfa, a favoured food plant for the lygus bug, succeeded in reducing the number of lygus bugs on the cotton. Populations of natural predators were also increased by providing them with extra food, a technique called **augmentation.** For example, the adult green lacewing (*Chrysopa carnea*) is attracted to the nutrient and so lays more eggs on the cotton. These hatch into carnivorous larvae which eat boll worms and so reduce their population density. Reduction in the use of pesticide also encouraged an increase in the number of predatory mites, small animals related to spiders, which also attack the boll worms. Previously, the mites were destroyed by the insecticides used to control the boll worm.

The key to this and all IPM programmes is careful field monitoring of pests and predator populations. This involves monitoring the cotton fields in the San Joaquin valley twice a week from mid-May until the end of August for the lygus bus. Additional monitoring for boll worm is needed from 1 August until mid-September. This carefully integrated programme of natural but augmented biological control coupled with the judicious use of insecticides has reduced the pest control costs of the San Joaquin cotton farmers by a factor of three since 1970.

The major problem with IPM is its initial development costs. However, the results, in terms of both a reduction in pesticide use and benefit–cost ratio, can be remarkable. For example, an IPM programme introduced in Indonesia to control the brown plant hopper (*Nilaparvata lugens*) a devastating insect pest of rice, resulted in:

• a nine-fold reduction in the amount of insecticide sprayed;

• a ten-fold reduction of costs;

• an increase in rice yields from 6 to 7.5 tonnes per hectare.

Clearly, IPM is the strategy for the future but much more research is needed before we see its wider implementation.

| QUESTION | 11.11 What do you consider to be the major benefits and major costs of IPM? |

| ANALYSIS | **Pulling the threads together**
This is a comprehension exercise.

A recurrent idea in this theme has been how all the various factors affecting crop productivity, soil, weather, pests and so on interact with each other. The following text describes the practices carried out on a farm north of Liverpool in England. Read the text carefully and then answer the questions at the end to see if you can apply the ideas you have read and learnt about in this theme.

The farm is on sandy soil but drainage is poor owing to the extraction of large quantities of sand for glass making. Tile drains were frequently blocked leading to waterlogging in the winter. A five-year crop rotation |

PEST CONTROL

of potatoes, winter wheat, winter barley, peas, winter wheat or barley was used. Up to 3.5 tons per acre of cow manure was applied prior to planting potatoes. This was a valuable crop which the farmer would have liked to grow more often in the same field. However, high cyst nematode populations prevented this. The potato variety used was Maris Piper which is resistant to cyst nematode A. Lime was applied every three years as ground limestone at a rate of 2 tons per acre.

A typical year in which winter wheat was grown started in early October with the land being ploughed and power harrowed to produce a seed bed. Sowing followed immediately to conserve moisture, with a phosphate/potash (potassium) fertiliser being applied either by broadcasting on to the seed bed or when the seed was drilled. A pre-emergence herbicide, Linuron, was applied immediately after drilling to control broad leaf weeds.

Germination typically occurred within two to three weeks with the plants overwintering at the 5 leaf stage in a mild winter. The crop was left until late February when, once soil temperatures exceeded 4°C, a nitrate fertiliser was applied. In late March the crop was harrowed and then rolled to encourage the plants to tiller.

The main application of nitrate fertiliser, usually as ammonium nitrate, occurred in April. If a warning was received from MAFF or ADAS the crop would be sprayed with a fungicide to control mildew and brown rust. If the crop was a little thin an extra broad leaf herbicide might have to be used at this stage.

Flag leaf and ear emergence occurred at the end of May/beginning of June. At this stage if more expensive milling wheat, used for making bread, was being grown, a fungicide would be applied to keep the ear clean with some extra ammonium nitrate added to boost the wheat's protein content. A straw-shortening hormone would be applied at the same time as the fungicide.

In July the crop would be sprayed against aphids, only if there was a warning from MAFF or ADAS. Fungicide, particularly against mildew, might also be applied. Harvesting occurred in August. Glyphosate would then be sprayed on to the field to provide a clean seed bed for the next crop. Ideally, the farmer liked to burn his cereal stubble but was now ploughing it in.

(a) Comment on the order of crops in this farmer's rotation.
(b) When and why was manure applied?
(c) What is the advantage of using autumn as opposed to spring sown cereals in the rotation?
(d) Why were phosphate and potash fertilisers applied in the autumn but nitrate fertiliser not applied until February once the soil temperature exceeded 4°C?
(e) What are the major pest control problems faced by this farmer? How did he control pests?
(f) Why were crops not sprayed unless warnings were given by MAFF or ADAS? Suggest the basis on which such warnings were given.
(g) What is the function of the straw shortener?
(h) How was the protein content of the wheat boosted? Explain the biochemical basis of this practice.
(i) How would you describe this farmer's tillage practice?
(j) Why did he need to use glyphosate?
(k) What is the function of stubble burning?

SUMMARY ASSIGNMENT

1. Write an essay on the biology of weeds, their effects on crop production and the methods used for their control.

2. How are pesticides applied to crops? What are the advantages and disadvantages of these different methods?

3. The government is considering banning the use of all pesticides and they have asked you as an applied ecologist to produce a report detailing the following points:
   (a) The types of pesticide in common use, their mode of action, toxicity and effects on humans.
   (b) The reasons why farmers use pesticides.
   (c) The possible adverse effects of pesticides.
   (d) The extent to which biological and cultural methods could be used to replace chemical control methods.

Theme 3

FRESHWATER POLLUTION

When Neil Armstrong stood on the moon and looked back to Earth his overriding impression was of a blue planet. Some 71% of the Earth's surface is covered with water. Essential to life, water is one of the most precious, but most abused, resources we have. Of particular concern is the way in which we pollute freshwater.

The words pollution and ecology have almost become synonymous. Certainly, both of them have become clichés. We are continually bombarded with news reports which tell us how polluted our environment has become. In particular, we need to understand:

- where pollutants come from and in what quantities;
- how pollutants produce their effects;
- what happens to pollutants once they are discharged into the environment;
- how we can monitor pollutant levels in the environment;
- how we can control the amount of a pollutant released into the environment.

We will explore these issues in this theme.

The over use of a beautiful area like the Norfolk Broads has led to the dramatic physical degradation seen in these two photographs. In addition chemical pollution, through run-off of nitrate fertilisers and sewage, has meant that many broads no longer support luxuriant and varied plant life.

Chapter 12

DIRTY WATER

LEARNING OBJECTIVES

After completing the work in this chapter you will be able to:

1. decide under what conditions freshwater can be said to be polluted;

2. describe the types, sources and effects of water pollution;

3. discuss the role of dilution, sedimentation, biodegradation and transformation in determining the fate of freshwater pollutants;

4. distinguish between biomagnification, bioaccumulation and bioconcentration.

12.1 WHAT IS POLLUTION?

An operational definition of freshwater pollution is:

The introduction by humans into the environment of substances or energy liable to cause hazards to human health, harm to living resources and ecological systems, damage to structure or amenity, or interference with legitimate uses of the environment.

In applying this definition it is important to realise that the levels of 'harmful substances' we can tolerate in the water will depend on the use to which we put that water. Water may be too polluted to drink but it may be satisfactory for industrial use. Water may be too polluted to swim in but it may be perfectly satisfactory for boating. One of the rivers near where I live is officially classified by the local water board as an effluent carrier. What constitutes pollution in such a water body? It is certainly heavily contaminated with treated sewage and industrial wastes but since it is not used for drinking water or any other human activity (that would be very unwise!) is it polluted?

Pollution is therefore a relative term; it relates to the use to which the water is going to be put. It is impossible to say this water is polluted unless we specify what the water is to be used for. Having defined use we can look up what the legal limits are for pollutants in such water. For example, Table 12.1 shows the limits set by one water board for drinking water.

ANALYSIS	**Defining pollution**
	This exercise involves the use of observation and discussion skills.

All of us have our own intuitive notion of what constitutes pollution. For most of us it is some physical degradation of our immediate environment. Look carefully at the photographs in Fig 12.1. Either on your own, or in small groups, try to rank the habitats shown in the photographs in order, from the most to the least polluted. Once you have done this write down the criteria you used to reach your decision.

Fig 12.1

You will find a brief description of each photograph on page 219. Read the descriptions. Does this change your rank order? To assess adequately how polluted a water body is we need to take into account its chemical and biological state, in addition to its appearance. We then need to interpret our findings in relation to the natural state of the water body and the use to which it is going to be put.

Table 12.1   Selected water quality criteria used by Anglian Water Authority for potable (drinking) water supply abstraction from rivers, fisheries and amenity and conservation.

| | | **Fisheries** | | |
	Potable water	**Salmonid**	**Cyprinid**	**High amenity and conservation**
temperature (°C)	30.0	22.0	28.0	25.0
dissolved oxygen (mg dm^{-3})	4.0	6.0	4.0	4.0
5-day BOD (mg dm^{-3})	9.0	6.0	9.0	9.0
anionic synthetic detergents (mg dm^{-3})	0.4	0.4	0.4	0.2
iron (mg dm^{-3})	3.0	1.5	1.5	1.0
manganese (mg dm^{-3})	0.15	1.5	1.5	1.0
zinc (mg dm^{-3})	7.5	0.75	3.0	2.0
copper (μg dm^{-3})	75.0	170.0	170.0	110.0
nickel (μg dm^{-3})	150.0	600.0	600.0	400.0
chromium (μg dm^{-3})	75.0	150.0	750.0	500.0
cadmium (μg dm^{-3})	7.5	2.3	2.3	1.5
mercury (μg dm^{-3})	1.5	0.23	0.23	0.15
lead (μg dm^{-3})	75.0	60.0	750.0	500.0
dissolved hydrocarbons (μg dm^{-3})	200.0			
total pesticides (μg dm^{-3})	2500.0			

Pure water

To judge adequately and objectively the effects of pollutants on freshwater ecosystems it is necessary that we first look at freshwater in its 'natural state'. Water is an excellent solvent so it naturally contains chemicals which it collects in its passage through the air, as water vapour, and through the **catchment** (the land area that supplies water, sediment and dissolved substances to a major river and its tributaries), after it has fallen as rain. Water can also carry large amounts of insoluble material in suspension, so-called **suspended solids**. The amount and types of impurities present will vary both with location and the time of year, and will determine the characteristics of a particular water body.

The chemical birth of freshwater

Once it is formed by evaporation, water vapour becomes contaminated with:

- **gases** such as oxygen, carbon dioxide, sulphur dioxide and various oxides of nitrogen (nitrous oxides). Dissolving carbon dioxide in water produces carbonic acid while sulphur dioxide and nitrous oxides produce sulphuric acid and nitric acid respectively. Thus rain water is naturally acidic with a pH of 5.2–6.5.

- **salts** derived from sea spray. When dissolved such salts will exist as ions, including chloride (Cl^-), sodium (Na^+), sulphate (SO_4^{2-}), magnesium (Mg^{2+}), calcium (Ca^{2+}) and potassium (K^+).

- **solid particles**, for example soil dust and pollen grains, which are being continually blown into the atmosphere. This material provides an additional source of ions. For example, rain falling over areas where the soil is derived from limestone (calcium carbonate), like East Anglia, contains high concentrations of calcium ions.

So, as it falls, rain is in fact a weakly acidic, dilute form of sea water. The composition of rainfall varies from place to place and it will also vary with time, the actual composition depending on the atmospheric conditions prevailing when the rain droplets were formed. As water runs off the land it collects further contaminants which are eventually carried into water bodies. These substances are derived from the rocks and the soils of the catchment. So differences in the geology and soils of catchments will therefore be reflected in the chemistry of the freshwater bodies they contain.

These differences in the natural chemistry of water from different catchments plays a crucial role in determining the impact of pollutants. For example, **acid rain** falling on to a catchment which has soils rich in bases, such as calcium carbonate, is neutralised. Consequently, there is no acidification of lakes in such catchments. But exactly the same rain falling on to soils derived from igneous rocks, which are deficient in bases, produces the acidified lakes seen in Sweden and southern Scotland (see Section 14.1). The hardness of water also has a direct effect on the toxicity of many pollutants. For example, heavy metals like lead are far less soluble in hard water than they are in soft water.

QUESTIONS

12.1 Consider a stream which is naturally poor in nutrients. It has few plants growing in it, few invertebrates and cannot provide enough food to support fish. A small amount of treated sewage is discharged into this stream. The decomposition of this sewage provides added minerals, stimulating the growth of plants. This increase in primary productivity leads to an increase in the number of invertebrates and there is now enough food for fish to survive. Is this stream being polluted? Justify your answer.

12.2 Would it be **(a)** necessary **(b)** cost effective **(c)** desirable to have water of the standard shown in Table 12.1 for use in an industrial plant?

12.2 TYPES, SOURCES AND EFFECTS OF FRESHWATER POLLUTANTS

Over 1500 substances have been listed as pollutants in freshwater. Biological, chemical and physical forms of water pollution can be classified into eight major types.

1. **Pathogenic organisms** (bacteria, viruses, protozoa and so on; Table 12.2 gives further details).
2. **Biodegradable organic wastes** (domestic sewage, animal manure and

Table 12.2 Pathogens in sewage and polluted streams.

Organism	Disease	Remarks
virus	poliomyelitis diaorrhea	exact mode of transmission not yet known; found in effluents from biological sewage purification plants
Vibrio cholerae	cholera	transmitted by sewage and polluted waters
Salmonella spp.	food poisoning	
Shigella	bacillary dysentery	polluted waters main source of infection
Leptospira ictero-haemorrhagiae	leptospirosis (Weil's disease)	carried by sewer rats
Entamoeba histolytica	dysentery	spread by contaminated waters and sludge used as fertiliser; common in warmer countries
Taenia spp.	tape worms	eggs very resistant; present in sewage sludge and sewage effluents; danger to cattle on sewage-irrigated land or land manured with sludge
Ascaris enterobius	nematode worms	danger to humans from sewage effluents and dried sludge used as fertiliser

trade wastes). Biodegradable materials are those organic substances which can be decomposed by microorganisms, usually bacteria and fungi, into inorganic substances. It is the aerobic decomposition of biodegradable organic compounds which creates the **biochemical oxygen demand** of a water body. Putting too much biodegradable material, for example sewage, into a water body can place too severe an oxygen demand on the water, leading to major pollution problems (see Chapter 13).

3. **Water-soluble inorganic compounds** (acids, salts, toxic metals and their compounds, anions, e.g. sulphide, sulphite and cyanide.)

4. **Plant nutrients** (water-soluble nitrate and phosphate salts). These may come from the breakdown of organic wastes, like sewage, or by leaching of fertilisers from fields, especially nitrates. In excess they cause **cultural eutrophication** (see Chapter 14).

5. **Insoluble and soluble organic chemicals** (oil, petrol, plastics, pesticides, solvents, PCBs, phenols, formaldehydes, plus a host more). This is the group which is causing the most concern at present since many are highly toxic, even at very low concentrations – less than 1 ppm.

6. **Suspended solids**. These are particles that are either insoluble or too large to be quickly dissolved. The tendency for suspended solids to settle to the bottom of a water body depends on their size, the flow rate and the amount of turbulence in the water body. Particles between 1 μm and 1 nm can remain suspended in the water indefinitely. These particles are called colloidal solids or simply colloids. Water which contains a lot of colloidal solids often has a milky appearance. The quantity of suspended solids affects the turbidity of the water; their quality affects the colour. **Turbidity** is a measure of the cloudiness of water due to the presence of suspended solids and colloids. The degree of turbidity is an important factor when studying water; pollution may cause a significant increase in turbidity, although pollutants may be present even in very clear water. In particular, all particles reduce light penetration into a water body, leading to a reduction in the rate of photosynthesis and hence plant growth. In addition, particles settling out on the bottom of, say, a stream may prevent some small invertebrates from living there, e.g. stone fly nymphs.

7. **Radioactive substances.**

Fig 12.2 The warm water effluent, which is polluting this waterway, is clearly visible in this photograph of a woodmill in Georgia, USA.

8. **Thermal pollution**, usually in the form of hot waste water from the cooling towers of power stations (Fig 12.2). This may raise the temperature of the receiving water, increasing the rate of decomposition of biodegradable organic wastes and reducing the water's capacity to hold oxygen (see Section 13.2).

Table 12.3 summarises the major sources, effects and methods for controlling these eight major types of water pollutant.

Table 12.3   Major water pollutants.

Pollutant	Sources	Effects	Control methods
disease-causing agents	domestic sewage; animal wastes	outbreaks of water-borne diseases, such as typhoid, infectious hepatitis, cholera and dysentery; infected livestock	treat waste water; minimise agricultural run-off; establish a dual water supply and waste disposal system
biodegradable oxygen-demanding wastes	natural run-off from land; human sewage; animal wastes; decaying plant life; industrial wastes (from oil refineries, paper mills, food processing, etc); urban storm run-off	decomposition by oxygen-consuming bacteria depletes dissolved oxygen in water; fish die or migrate away; plant life destroyed; foul odours; poisoned livestock	treat waste water; minimise agricultural run-off
Inorganic chemicals and minerals			
acids	mine drainage; industrial wastes; acid deposition e.g. acid rain	kills some organisms; increases solubility of some harmful metals, e.g. aluminium	seal mines; treat waste water; reduce atmospheric emissions of sulphur and nitrogen oxides
salts, e.g. sodium chloride	natural run-off from land; irrigation; mining; industrial wastes; oil fields; urban storm run-off; de-icing of roads with salts	kills freshwater organisms; causes salinity build-up in soil; makes water unfit for domestic use, irrigation and many industrial uses	treat waste water; reclaim mined land; use drip irrigation; ban brine effluents from oil fields
mercury	natural evaporation and dissolving; industrial wastes; fungicides	highly toxic to humans (especially methyl mercury)	treat waste water; ban inessential uses

Table 12.3   **Major water pollutants** (*continued*).

Pollutant	Sources	Effects	Control methods
Plant nutrients (phosphates and nitrates)	natural run-off from land; agricultural run-off; mining; domestic sewage; industrial wastes; inadequate waste water treatment; food-processing industries; phosphates in detergents	algal blooms and excessive aquatic growth; kills fish and upsets aquatic ecosystems; eutrophication; possibly toxic to babies and livestock; foul odours	advanced treatment of industrial, domestic and food-processing wastes; recycle sewage and animal wastes to land; minimise soil erosion
Organic chemicals oil and grease	machine and automobile wastes; pipeline breaks; offshore oil well blow-outs; natural ocean seepages; tanker spills and cleaning operation	potential disruption of ecosystems; economic, recreational and aesthetic damage to coasts, fish and waterfowl; taste and odour problems	strictly regulate oil drilling, transportation and storage; collect and reprocess oil and grease from service stations and industry; develop means to contain and mop up spills
pesticides and herbicides	agriculture; forestry; pest control	toxic or harmful to some fish, shellfish, predatory birds and mammals; concentrates in human fat; some compounds toxic to humans; possible birth and genetic defects and cancer	reduce use; ban harmful chemicals; switch to biological and cultural control of pests
plastics	homes and industries	kills fish; effects mostly unknown	ban dumping, encourage recycling of plastics; reduce use in packaging
detergents (phosphates)	homes and industries	encourages growth of algae and aquatic weeds; kills fish and causes foul odours as dissolved oxygen is depleted	ban use of phosphate detergents in crucial areas; treat waste water
chlorine compounds	water disinfection with chlorine; paper and other industries (bleaching)	sometimes fatal to plankton and fish; foul tastes and odours; possible cancer in humans	treat waste water; use ozone for disinfection and activated charcoal to remove synthetic organic compounds
suspended solids	sewage and trade wastes; natural erosion, poor soil conservation; run-off from agricultural, mining, forestry and construction activities	major source of pollution; fills in waterways, harbours and reservoirs; reduces shellfish and fish populations; reduces ability of water to assimilate oxygen-demanding wastes	treat waste water; more extensive soil conservation practices
heat	cooling water from industrial and electric power plants	decreases solubility of oxygen in water; can kill some fish; increases susceptibility of some aquatic organisms to parasites, disease and chemical toxins; changes composition of and disrupts aquatic ecosystems	decrease energy use and waste; return heated water to ponds or canals or transfer waste heat to the air; use to heat homes, buildings and greenhouses
radioactive substances	natural sources (rocks and soils); uranium mining and processing; nuclear power generation; nuclear weapons testing	cancer; genetic defects	ban or reduce use of nuclear power plants and weapons testing; more strict control over processing, shipping, and use of nuclear fuels and wastes

DIRTY WATER   **171**

Point and non-point sources

If we want to control the release of freshwater pollutants we need to consider the route by which a pollutant reaches a water body. For this purpose it is useful to distinguish between **point** and **non-point** sources (Fig 12.3). A point source is a source that discharges effluents, such as waste water, through pipes, ditches and sewers into bodies of water at specific locations. Examples include factories, sewage works and power stations. A non-point source consists of many widely scattered sources discharging effluents into a water body over a wide area. Examples would include run-off into surface and ground water from agricultural land, car parks and streets.

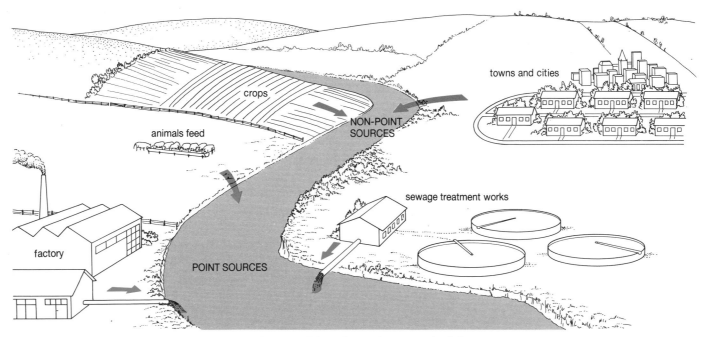

Fig 12.3 Point and non-point sources of pollution.

Point and non-point sources tend to produce quite different patterns of discharge. Point sources usually produce regular discharges, year round. Significant amounts of pollution from non-point sources may only occur as a result of storms which, for example, wash out drains and cause storage tanks at sewage works to overflow.

QUESTIONS

12.3 The concentration of freshwater pollutants is often measured in parts per million (ppm) or parts per billion (ppb). 1 mg (10^{-3} g) of a pollutant in 1000 g (1 dm^3) of water = 1 ppm. 1 µg (10^{-6} g) of a pollutant in 1 dm^3 of water = 1 ppb. Express the concentrations of fluoride, nickel, lead and manganese given in Table 12.1 in either ppm or ppb.

12.4 Do you think it will be easier to regulate the release of pollutants from point or non-point sources? Explain your answer.

12.5 (a) Why will rivers which are in flood appear very turbid?
   (b) Why will rivers draining clay beds appear more turbid than those draining rocky areas?

DIRTY WATER

12.3 THE FATE OF FRESHWATER POLLUTANTS

Given a sample of an effluent, we could in theory determine the composition and concentration of the chemicals it contains. However, the concentration and chemical form of pollutants change once they are discharged into a surface water body as a result of four natural processes:

1. dilution;
2. sedimentation;
3. biodegradation;
4. chemical transformation, which may occur inside an organism.

We must be aware of these changes if we are going to predict the ecological impact of discharging a pollutant.

Dilution

The amount of dilution of all pollutants depends on the volume and flow rate of a water body. In a large, rapidly flowing river even quite large amounts of a pollutant can become rapidly diluted to insignificant levels, causing no problem. However, the same pollutant discharged into a small stream or lake could have catastrophic effects. It is worth remembering that dilution will be greatly reduced during dry spells or if large amounts of water are being withdrawn from a river for cooling or drinking.

Dilution is not the answer to all our problems. Effective dilution relies on mixing. So lakes which are stratified (Fig 12.4) will be less effective at diluting a pollutant since the stratified layers undergo little mixing. Furthermore, some compounds, particularly synthetic, organic molecules like pesticides, can be toxic at even very low concentrations. Nonetheless, dilution remains an important tool for controlling pollution, particularly for largely biodegradable effluents like sewage.

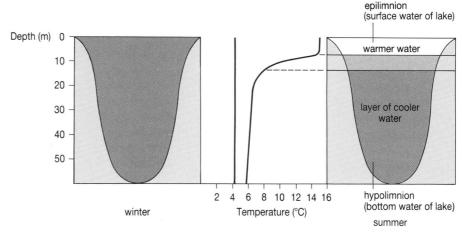

Fig 12.4 Thermal stratification of a lake in the summer months reduces mixing of the water which may exacerbate pollution problems.

Sedimentation

Effluents like sewage contain large amounts of suspended solids. Such material will naturally fall to the bottom of a water body under the appropriate conditions of flow and turbulence. This can of course cause problems as the sediment may, for example, prevent photosynthesis by blocking the penetration of light into the water. Nonetheless, at some point below a discharge a river will have rid itself of the additional suspended solid material added by the effluent through sedimentation of suspended material.

Sedimentation can also remove trace amounts of some organic and inorganic pollutants, which become attached to particles that settle and accumulate in the mud at the bottom of lakes and slow flowing rivers. However, such sediment can represent a potential time bomb since the pollutants can be released back into the water. This may happen when the mud is stirred up by dredging or by chemical changes in potential pollutants brought about by changes in the chemistry of the water. For example, heavy metals like cadmium and lead 'immobilised' in bottom sediments can be released by increased salinity, increased acidity and changes in **redox** conditions. (A medium, water or sediment, low in oxygen is a reducing environment. The degree of reduction can be measured as the **redox potential**. Well oxygenated waters have redox potentials greater than +500 mv. Anoxic conditions are indicated by redox potentials of about +200 mv.) There is evidence that copper, zinc and cadmium are released from anoxic (containing no oxygen) sediments into surface waters in the Mersey estuary when there is a particularly low tide, which prolongs the exposure of the sediments to the air and so changes their redox potential.

Biodegradation

Animals and microbes will use biodegradable organic material, carbohydrates, proteins and so on, as food. As they oxidise the material they will convert it to carbon dioxide, nitrates and phosphates. A general equation for this process would be:

$$\underset{\substack{\text{carbohydrates} \\ \text{and proteins}}}{C, H, O, N, P, S} \dots O_2 \xrightarrow{\text{aerobes}} \underset{\text{carbon}}{CO_2} + \underset{\text{water}}{H_2O} + \underset{\text{nitrates}}{NO_3^-} +$$

$$\underset{\text{phosphates}}{PO_4^{3-}} + \underset{\text{sulphates}}{SO_4^{2-}} + \text{new cells} + \text{energy}$$

In the absence of dissolved oxygen, aerobic bacterial processes stop. Under these conditions anaerobic bacteria make use of the oxidising potential of inorganic salts, like nitrates, sulphates and phosphates, to convert organic material to carbon dioxide and methane, so providing energy. A generalised equation for anaerobic bacterial activity is:

$$\underset{\substack{\text{carbohydrates} \\ \text{and protein}}}{C, H, O, N, P, S} + \underset{\text{nitrates}}{NO_3^-} + \underset{\text{phosphates}}{PO_4^{3-}} + \underset{\text{sulphates}}{SO_4^{2-}} \xrightarrow{\text{anaerobes}}$$

$$\underset{\substack{\text{carbon} \\ \text{dioxide}}}{CO_2} + \underset{\text{methane}}{CH_4} + \underset{\text{nitrogen}}{N_2} + \underset{\text{phosphine}}{PH_3} + \underset{\substack{\text{hydrogen} \\ \text{sulphide}}}{H_2S} + \text{new cells} + \text{energy}$$

This biodegradation may cause serious pollution problems, as discussed in Chapter 13, but it does, ultimately, remove the pollutant from the water. Biodegradation is of course totally ineffective in removing non-biodegradable and persistent pollutants, like the insecticide DDT, which are stored in an organism's body, rather than degraded.

Chemical transformation

An effluent being discharged into a water body is likely to be a complex mixture of chemicals. Not surprisingly, once they are discharged these chemicals will interact with each other and the environment, so changing their chemical nature. Thus changes in pH, redox potential and temperature can radically alter the chemical nature of an effluent. This presents a real headache for pollution scientists. We may be able to characterise the chemical nature of the effluent before it is discharged but once it is in the environment it may change its character quite surprisingly, with potentially disastrous consequences for aquatic organisms. For example, if an effluent containing soluble salts of iron or aluminium is discharged into a naturally alkaline river then the metals can precipitate out as aluminium or iron hydroxides. In the case of iron a clear effluent discharged into a clear alkaline stream can produce a bright orange colouration as iron hydroxides form. Although neither aluminium or iron hydroxide are toxic the precipitate prevents light penetration and so inhibits plant growth.

Biotransformation

Living organisms provide an ideal environment in which chemical transformations can occur. An obvious example is the oxidation of biodegradable organic compounds to carbon dioxide and water during respiration. In addition, organisms can convert non-toxic compounds into highly toxic ones or they can accumulate compounds to levels where they become toxic. For example, metallic mercury, which is not particularly toxic if swallowed, is extremely toxic if it is converted to methyl mercury (Fig 12.5). This occurs in aquatic ecosystems as a result of the biotransformation of inorganic mercury by bacteria and fungi. For example, anaerobic bacteria, like *Clostridium*, living in anoxic sediments at the bottom of lakes, aerobic fungi, like *Neurospora*, and aerobic bacteria, e.g. *Pseudomonas*, can all methylate mercury, particularly under acidic conditions. Methyl mercury killed

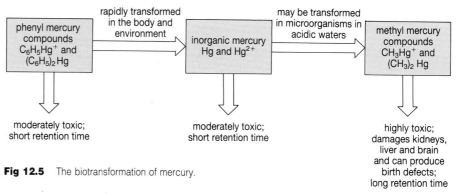

Fig 12.5 The biotransformation of mercury.

649 people and poisoned 1385 people from the Japanese fishing village of Minimata in the late 1950s. In 1972, 459 Iraqis died as a result of eating wheat seed which had been coated with methyl mercury to inhibit the growth of fungi.

Biomagnification

You are probably familiar with the picture shown in Fig 12.6. This process is called **biomagnification** – an increase in the concentration of a pollutant

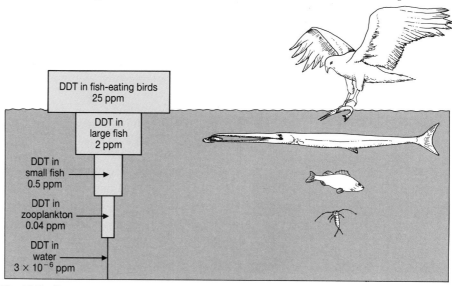

Fig 12.6 Concentration of DDT in an aquatic food chain.

in animal tissue in successive members of a food chain. The increase in the concentration of a chemical as we move up the food chain could be occurring by two distinct mechanisms.

1. **Bioconcentration** – the increase of pollutant concentration from water when passing directly into an aquatic species.
2. **Bioaccumulation** – the increase of pollutant concentration as a result of the combined intake from food as well as from water.

The available evidence suggests that intake from food is unlikely to be the major source of the residues of persistent organic pollutants, like DDT, in aquatic organisms. It could be that organisms higher up food chains bioconcentrate more of these fat-soluble organic compounds because they contain more fat. Again, the evidence is equivocal. Indeed, many studies have now shown that persistent pollutants sometimes do, and sometimes do not, increase in concentration along aquatic food chains.

ANALYSIS

A complex problem

This exercise will help you to appreciate the complex nature of pollution.

During a particularly hot, dry summer, fishermen noticed that the trout in their local lake were dying. The dead fish were collected and sent to a local laboratory for analysis. They were found to contain high levels of dimethyl mercury, cadmium and the banned pesticide dieldrin. Scientists from the laboratory were sent to the lake to carry out an extensive survey. They collected water, sediment, and phytoplankton and zooplankton samples, which were then analysed for the pesticide dieldrin. In addition, they analysed the levels of cadmium in the water during the summer and later in the same year in the winter when they found the cadmium concentration in the water had increased significantly.

Historical research indicated that the lake had once been used as a dump for wastes from a local electrical factory where both cadmium and mercury had been used. Local farmers had once used dieldrin as a seed dressing to protect winter wheat from wheat bulb fly. However, it was also noted that this was the first time that fish had ever died in such numbers.

(a) Account for the presence of dieldrin in the lake.
(b) Explain the increase in the level of cadmium in the water during the winter.
(c) Account for the presence of dimethyl mercury in the fish.
(d) The concentrations of dieldrin, dimethyl mercury and cadmium in the fish were, individually, below lethal levels. No other contaminants were found in the fish. Suggest why the fish might have died.

QUESTIONS

12.6 (a) Why would introducing small quantities of treated sewage into a large, fast flowing river be unlikely to have any detrimental effect?
(b) How could introducing the same amount of sewage into a small, clear stream result in a water body which is both turbid and smells of rotten eggs (hydrogen sulphide)?

12.7 (a) How can pollutants be transfromed into new chemical species in freshwater ecosystems?
(b) Why is such transformation a problem for scientists investigating the effects of introducing a pollutant into a freshwater ecosystem?

12.4 CASE STUDY: SEWAGE

Pollutants originate from domestic, agricultural and industrial sources. Usually, they do not come in a pure form. Rather the effluents derived from these sources are a complex cocktail of pollutants. To illustrate this complexity, which presents major problems for regulatory bodies, let us examine the composition of a common effluent, sewage.

It must be stressed that, apart from cities which are sited on estuaries, e.g. Liverpool, little river pollution is due to the discharge of raw (untreated) sewage. Rather river pollution from sewage is largely the result of discharging inadequately treated sewage effluents contaminated with industrial wastes or from the overflow of sewers. This is not true of marine discharges where raw sewage is still pumped into the sea. For example, in southern Cornwall there are 59 major sea outfall pipes, 29 of which release raw sewage, leading to polluted sea water and beaches, as occurs in Penzance Bay.

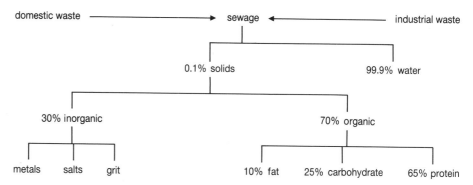

Fig 12.7   The composition of sewage. Note that the major component is water.

The composition of raw sewage is shown in Fig 12.7. Note that it consists of both domestic sewage (human faeces and urine, waste water from sinks and baths) and industrial sewage (the waste water from factories, abbatoirs and hospitals). Sewage is therefore a complex mixture of organic and inorganic substances but the main constituent is water.

Suspended organic materials in raw sewage consist mainly of fats, proteins, carbohydrates (including starches and celluloses), organic acids, soaps and detergents. The dissolved organic material consists largely of sugars, organic acids, including amino acids, and detergents. Inorganic constituents include anions (cyanide, sulphide and chloride), heavy metals (cadmium and lead) and grit.

Raw sewage also contains a large number of living organisms, some of which may be pathogenic (see Table 12.2). While chlorination and filtration of drinking water has removed the threat posed by such pathogens in more developed countries, like Britain, they still represent a major problem in less well developed countries. Even in Britain, drinking water is not necessarily pathogen free as shown by an outbreak of diarrhoea in the Swindon area of southern Britain in 1989. This was caused by *Cryptosporidium*, a bacterium from livestock which is not killed by chlorination, entering the water supply.

In a sewage works the raw sewage is treated, largely to reduce the amount of biodegradable organic material and suspended solids like grit. The number of living organisms is also greatly reduced. The sewage effluent is now released to a stream near the sewage works at rates of only a few per cent of the total stream discharge. At such a dilution the natural stream community will decompose the remaining organic material without severe problems. However, while sewage treatment reduces the amount of organic material in the effluent discharged from the sewage works it still contains a large amount of dissolved inorganic materials such as nitrates and phosphates, which can lead to eutrophication (see Section 14.2). In addition, the effluent may also contain toxic chemicals, such as heavy metals, phenols and pesticides. These may be present at extremely low concentrations, less than 1 ppm, but can still have a devastating effect on aquatic life (see Section 14.3). Sewage treatment is dealt with in more detail in another book in this series by Jane Taylor, *Micro-organisms and Biotechnology*.

SUMMARY ASSIGNMENT

1. Why is it difficult to give an objective definition of pollution?

2. Using Table 12.3 as a basis prepare a display showing the main types, sources and effects of water pollution. Collect and include newspaper cuttings, photographs, drawings and graphs to stress the relative importance of different types of freshwater pollution.

3. Explain why dilution and sedimentation are not always the solution to water pollution problems. Give examples and conditions when these solutions are, and are not, applicable.

4. Distinguish between and give examples of: **(a)** point and non-point sources of pollution **(b)** biodegradation, biomagnification, bioaccumulation and bioconcentration.

Chapter 13

ORGANIC POLLUTION

Organic matter is a major freshwater pollutant. Organic material may be biodegradable, inert (non-biodegradable) or toxic. It may produce foam, unpleasant tastes and odours, or interact with other pollutants to produce toxic compounds, e.g. chlorophenols. The major polluting effects of organic materials result from the fact that many are biodegradable. This means that bacteria and other organisms can break them down into simpler organic and inorganic compounds, using up oxygen in the process. This chapter concentrates on the consequences of this consumption of dissolved oxygen, the **oxygen demand**, on the flora and fauna of rivers.

LEARNING OBJECTIVES

After completing the work in this chapter you will be able to:

1. list some sources of oxygen-demanding pollutants;

2. explain why biodegradable organic pollutants reduce the dissolved oxygen content of a water body;

3. use biochemical oxygen demand and population equivalents to compare the 'strength' of different effluents;

4. describe and account for the characteristic changes which occur in the dissolved oxygen content, and the flora and fauna in rivers being polluted by biodegradable organic material.

13.1 SOURCES OF BIODEGRADABLE ORGANIC POLLUTANTS

Fig 13.1 This plastic lined pit is trapping silage liquor which has run off from a silage clamp. If this highly corrosive and biodegradable material were to enter a watercourse then it would have a disastrous effect on aquatic life.

Sources of biodegradable organic pollutants (BOPs) include:

- run-off from land and roads, e.g. dog faeces;
- sewage which includes both human and industrial wastes;
- farm wastes, e.g. slurry (animal faeces, urine), silage liquor (Fig 13.1);
- industrial wastes from oil refineries, paper mills, food processing and pharmaceutical factories.

All these sources produce effluents which are rich in easily oxidised organic compounds like carbohydrates (sugars, starch and so on), proteins, lipids and nucleic acids. This organic material is rapidly colonised and decomposed by bacteria and fungi. As aerobic decomposition requires oxygen the decomposing effluent exerts an **oxygen demand** on the water into which it is discharged. If not too much effluent is discharged, if it is well diluted and if the rate at which oxygen dissolves in the receiving water is sufficiently high the effluent will cause little damage. However, if the oxygen demand of the effluent exceeds the oxygen supply then the receiving water will turn **anoxic** (= no oxygen). Under these circumstances fish die, the aquatic community changes and anaerobic decomposition processes predominate. The products of such anaerobic decay, including hydrogen sulphide, produce the foul odours characteristic of anoxic water bodies.

We can use the **biochemical oxygen demand (BOD)** test to compare the

Table 13.1 Comparative strengths of liquid waste from industry

Type of waste producer	5-day BOD (g m^{-3})	Population equivalent per m^3 of waste	pH value	Suspended solids (g m^{-3})
cotton mill	200–1000	3.3–16.7	8–12	200
tannery	1000–2000	16.7–33.3	11–12	2000–3000
brewery	850	14.2	4–6	90
dairy	600–1000	10–16.7	acid	200–400
slaughterhouse	1500–2500	24–41.7	7	800
grass silage	50 000	833.3	acid	low
farm	1000–2000	16.7–33.3	7.5–8.5	1500–3000
poultry	500–800	8.3–13.3	6.5–9	450–800

strengths of different effluents. BOD is dealt with in detail in Chapter 15. Briefly, it is a measure of the polluting capacity of an effluent due to the dissolved oxygen taken up by microorganisms as they decompose the organic matter contained in the effluent. The greater the effluent's BOD the greater its potential for reducing the level of dissolved oxygen in a water body. Typical values for domestic sewage are 250–350 g m^{-3}. By comparison, clean river water normally has a value of 3 g m^{-3} or less, while a very polluted stream might have a value of around 10 g m^{-3}. The 'strengths' of different effluents are shown in Table 13.1.

Population equivalents represent a way of comparing the strengths of oxygen-demanding wastes by comparing them to the BOD of domestic sewage. An individual contributes an average BOD value of about 60 g m^{-3} per day. Therefore a daily discharge of an industrial effluent with a BOD of 300 g m^{-3} would have a population equivalent of 5.

While BOD and population equivalents are useful indicators of the strength of an industrial effluent they only relate to the effects of biodegradable organic matter. Remember, industrial effluents may pollute in other ways, including:

- altering the pH of the receiving water;
- adding toxic materials such as heavy metals and organic compounds like phenol;
- adding hot water to the water body.

QUESTIONS

13.1 (a) What is the oxygen demand of an effluent?
   (b) An industrial effluent has a BOD of 2400 g m^{-3}. What is its population equivalent?
   (c) Which type of effluent in Table 13.1 exerts the greatest BOD?

13.2 During the summer the deep pond in my village gives off gases like hydrogen sulphide (which smells like rotten eggs) but does not do so in the winter. Explain this observation.

13.3 (a) What would happen to a river, its animals and plants if the water's BOD exceeded the rate at which oxygen was dissolving in the water?
   (b) How might introducing a weir, an artificial 'waterfall', into the stream help to remedy the situation?

13.2 USING UP OXYGEN

A minimum of 5 g m^{-3} of dissolved oxygen is considered necessary to support a rich and balanced aquatic community. Since the **saturation value** (the amount of oxygen which can be dissolved in a cubic metre of oxygen at a given temperature) at 15°C is only 9.8 gm^{-3} it is apparent that any factor which reduces the oxygen concentration, even slightly, may affect aquatic organisms. As aerobic aquatic organisms respire they consume oxygen, exerting a BOD on the water body. In addition, inorganic substances like sulphite (SO_3^{2-}), sulphide (S^{2-}) and iron (Fe^{2+}), as well as easily oxidised organic compounds like oxalic acid, take part in reactions which consume oxygen and so exert a **chemical oxygen demand (COD)** on the water body.

Replenishing the supply of dissolved oxygen

In studying the effects of pollutants on freshwater ecosystems, we need to consider not only how much oxygen is dissolved in the water, but also how quickly oxygen will dissolve into a water body. This depends on three factors.

1. The amount of oxygen already dissolved in the water. This will determine the rate at which oxygen diffuses from the atmosphere into the water.
2. Turbulence will increase the rate at which oxygen is transferred from the air into the water. Turbulent flow mixes the surface waters, where the oxygen has just dissolved, with the main body of water so reducing the oxygen concentration in the surface waters. Rapidly flowing turbulent streams can therefore take up oxygen more rapidly than smoothly flowing slow ones.
3. The surface area to volume ratio of a water body. Wide, shallow water bodies will have greater surface area to volume ratios than deep, narrow water bodies so oxygen will diffuse faster into the former than the latter. Turbulence will also increase the surface area to volume ratio of a water body.

In general, small, rapidly flowing streams are nearly always saturated with dissolved oxygen; in contrast, large, sluggish rivers may have concentrations as low as 50% of that required for saturation. The situation in lakes is rather more complex since they can become stratified during the summer months (see Fig 12.4). This means that there is little transfer of oxygen from the surface to the bottom of the lake. As a result the hypolimnion often becomes anoxic (contains no oxygen) in the summer. This has major implications for discharging effluents which increase the BOD of water, for example sewage, into lakes.

ANALYSIS	**Dissolved oxygen and temperature**

Dissolved oxygen and temperature
This exercise will give you practice in drawing and interpreting graphs.

(a) Plot the data in Table 13.2 on a suitable graph.
(b) Describe the relationship between dissolved oxygen and temperature.
(c) Does the nature of the relationship surprise you?
(d) Power stations often discharge warm water from their cooling towers into rivers. What are the implications of this 'thermal pollution' for the organisms which inhabit the river?
(e) An industrialist wishes to discharge an effluent at 45°C which contains a lot of sulphide into a small stream. Predict and explain the possible effects of such a discharge.

Oxygen sag curves

Imagine we are investigating the effect of discharging sewage into a stream. We collect samples at regular intervals below the point of discharge and determine the amount of dissolved oxygen in the samples using an

Table 13.2 Solubility of oxygen in water – saturation concentrations.

Temperature (°C)	Dissolved oxygen (g m⁻³)	Temperature (°C)	Dissolved oxygen (g m⁻³)
0	1.42	16	9.6
1	13.8	17	9.4
2	13.4	18	9.2
3	13.1	19	9.0
4	12.7	20	8.8
5	12.4	21	8.7
6	12.1	22	8.5
7	11.8	23	8.4
8	11.5	24	8.3
9	11.2	25	8.1
10	10.9	26	8.0
11	10.7	27	7.9
12	10.4	28	7.8
13	10.2	29	7.6
14	10.0	30	7.5
15	9.8		

Fig 13.2 Measuring oxygen levels in experimental ponds. The oxygen probe is submerged in the water and is sending electrical signals back to the meter from which oxygen concentration can then be read.

oxygen meter and probe (Fig 13.2). Plotting the amount of dissolved oxygen in the water against the distance downstream from the point of discharge gives a characteristic oxygen sag curve (Fig 13.3). The exact shape of this curve will depend on the following factors.

- The oxygen demand of the effluent and the receiving water. The greater the oxygen demand of the effluent the greater the reduction in the amount of dissolved oxygen.

- The water flow rate and the amount of turbulence. The greater the flow rate and volume of the receiving water body the greater the dilution of the pollutant and the faster it is swept away. Since fast flowing turbulent rivers also contain more oxygen the rate of oxidation will also be high. However, dilution and oxidation will be greatly reduced if the flow is decreased, either by dry weather, or the withdrawal of large quantities of water for irrigation or industrial or domestic purposes.

- The type of effluent. For example, sewage already contains a large microbial community and so it is broken down quickly. However, effluents which do not contain endogenous microbial communities, for example those containing phenol, a powerful disinfectant, will take longer to decompose, since it takes time for the microbial flora to colonise and, subsequently, develop on the effluent. Other effluents may contain highly refractory (difficult to decompose) materials. For example, pulp mills discharge effluents rich in cellulose, which is difficult to degrade.

- The temperature of the water. Warm water contains less dissolved oxygen than cold water, while the warmer the water the faster the rate of bacterial decomposition of BOPs. These factors will further compound the problem of adding BOPs to rivers during dry, warm weather when stream flow and turbulence are low.

- In addition to their organic content, many low quality effluents, such as poorly treated sewage, will contain substantial amounts of ammonia (NH_3) produced, for example, by the decomposition of proteins. In addition to being directly toxic to fish and other animals, ammonia further increases the deoxygenation of the receiving water as it is oxidised first to nitrite and then to nitrate both processes which consume oxygen.

ORGANIC POLLUTION

13.4 (a) In Fig 13.3 the point of minimum dissolved oxygen occurs some distance below the point of discharge. Account for this observation.

(b) What effect would discharging an effluent with a higher BOD have on the shape of the oxygen sag curve?

(c) Why does the level of dissolved oxygen in the river eventually return to its original level?

(d) What would the effect be on the levels of dissolved oxygen in the river and the shape of the oxygen sag curve if, at point A in Fig 13.3, (i) more sewage with a high BOD was introduced; (ii) hot water from a power station was introduced; (iii) water was abstracted for irrigation? Explain your answer in each case.

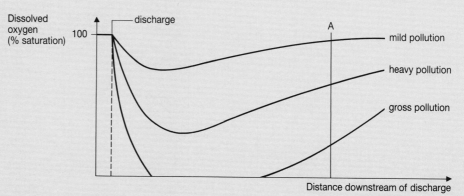

Fig 13.3 Oxygen sag curves. The degree of oxygen depletion depends on the level of pollution. The line A is referred to in Question 13.4.

13.5 (a) Why will adding sewage to a lake potentially have a more damaging effect than adding the same effluent to a rapidly flowing stream?

(b) Why would it be sensible to install small weirs in a stream receiving a discharge from a sewage works?

(c) Give three reasons why sewage discharged into a stream during the summer could have a greater impact on the dissolved oxygen content of the water than the same sewage discharged into the same stream during the winter.

13.3 THE EFFECT OF BOPS ON PLANTS AND ANIMALS

Changes in the amount of dissolved oxygen in a river below the point of discharge produce characteristic changes in the type and diversity of aquatic organisms. These are shown in Fig 13.4 which also summarises the physical and chemical changes occurring in the receiving water. Note that sewage, in particular, often contains a lot of suspended organic material and grit which increases the turbidity of the water. This prevents the penetration of light and so further inhibits the growth of autotrophs like algae in the decomposition and septic zones.

To account for the floral and faunal changes shown in Fig 13.4 let us take a trip down an imaginary river which is being polluted by oxygen-demanding wastes from, say, a sewage works. As we go we will collect samples of water for analysis and take samples of the fauna and flora.

Clean water zone

Above the point of discharge (or outfall), shown in red in Fig 13.4, we will assume that we have an unpolluted stream. Here there is a great diversity of animals and plants but no species or type is dominant.

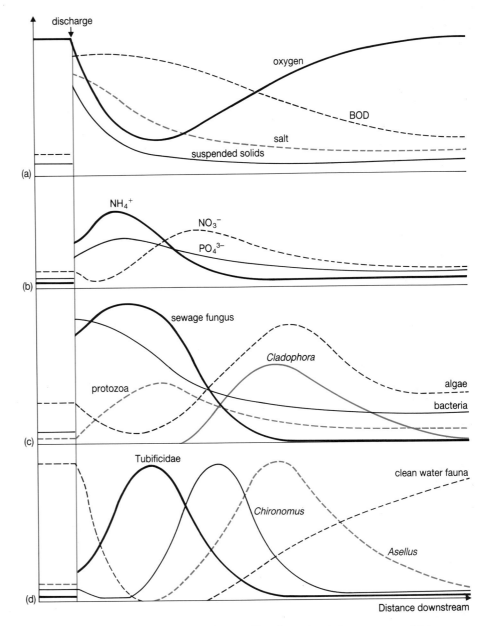

Fig 13.4 Changes in the water quality and communities of organisms below a sewage outlet.
(a) Physical changes (The presence of salt indicates sewage pollution.) **(b)** chemical changes **(c)** changes in microorganisms **(d)** changes in macro invertebrates.

Point of discharge

The introduction of an oxygen-demanding pollutant, such as sewage, starts a succession of fairly well organised events including (1) a decrease in the number of species present; (2) an increase in the number of individuals of a pollution-tolerant species. These factors lead to a decrease in **species diversity.**

Zone of degradation

The pollutant is first physically mixed into the receiving water which suspends material in the discharge, for example grit and fine organic particles, so reducing light availability to plants and smothering animals. Bacteria and other saprophytic microorganisms, like fungi, colonise the

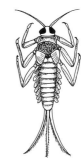

Stonefly nymph: brownish colour, crawls very slowly, no gills on abdomen, two tail processes.

Mayfly nymph: brown mottled colour, gills along abdomen, herbivorous, swims by beating abdomen up and down, three tail processes.

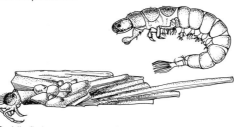

Caddis-fly larvae: some species encase themselves with small pebbles or plant material while others are caseless, the head is often brightly coloured.

Gammarus: freshwater shrimp, grey–brown colour, body flattened sideways, feeds on dead and decaying matter, swims on its side.

Fig 13.5 Some typical representatives of the clean water fauna.

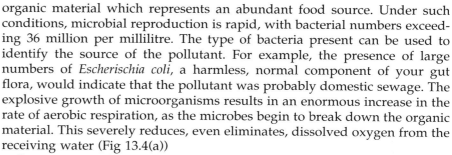

Fig 13.6 Larva of bee fly or rat-tail maggot. The animal breathes through its 'tail' which is telescopic and can be extended a considerable distance.

organic material which represents an abundant food source. Under such conditions, microbial reproduction is rapid, with bacterial numbers exceeding 36 million per millilitre. The type of bacteria present can be used to identify the source of the pollutant. For example, the presence of large numbers of *Escherischia coli*, a harmless, normal component of your gut flora, would indicate that the pollutant was probably domestic sewage. The explosive growth of microorganisms results in an enormous increase in the rate of aerobic respiration, as the microbes begin to break down the organic material. This severely reduces, even eliminates, dissolved oxygen from the receiving water (Fig 13.4(a))

Organisms intolerant of low levels of dissolved oxygen (the clean water fauna of Fig 13.4(d)), e.g. fish, stoneflies, mayflies, caddis flies and *Gammarus* (Fig 13.5) disappear while tolerant organisms, especially scavengers like the larvae of hover flies (Fig 13.6), increase in number. Overall, then, the number of different species of an organism in the community falls but numbers of individuals of tolerant types may increase. In other words, the **species diversity** in this zone is much less than that of the clean water zone.

Septic zone

The exact point at which the septic zone begins, if one occurs, varies with season and other factors, for example water temperature. If the oxygen demand does exceed supply then the water and sediments become anoxic. The lack of free dissolved oxygen in the water kills many aerobic microorganisms and nearly all larger plants and animals. This further increases the supply of biodegradable organic material in the water. The anoxic conditions favour facultative and obligate anaerobes, and gases, such as hydrogen sulphide and methane, may be evolved as a result of anaerobic respiration, resulting in foul odours. If heavy metals are present in the water, for example iron, these may react with hydrogen sulphide to form insoluble sulphides which colour the water black.

ORGANIC POLLUTION

Fig 13.7 Sewage fungus being carried down a stream on a piece of raw sewage. The fungus will then grow into the characteristic matts of white threads which will line the edge of the stream.

If the pollution is particularly bad then the water and banks in this region may hang with the white threads of **'sewage fungus'** (Fig 13.7) Despite its name, sewage fungus is composed mainly of bacteria not fungi. The white threads consist of the aerobic bacterium *Sphaerotilus natans*, which in turn provides a matrix in which other bacteria, diatoms and protozoa can grow. The whole assemblage constitutes sewage fungus. While this community of organisms can grow at low oxygen concentrations it cannot do so in totally anoxic waters. Sewage fungus has a high demand for oxygen.

The number of different species reaches a minimum in the septic zone while the number of individuals of tolerant species may reach a maximum. So this zone has the lowest species diversity of all.

Recovery zone

The septic zone gradually merges into the recovery zone. By now, most of the organic matter has been decomposed. Dissolved oxygen levels start to rise as oxygen diffuses into the water from the air faster than it is consumed by microbial respiration. This process is accelerated by the use of weirs which increase the turbulence. As a result, aerobic organisms start to reappear.

Since the amount of suspended material is now greatly reduced, light can once again penetrate the water column. Simultaneously, the availability of mineral nutrients (e.g. nitrate and phosphate) increases as the result of microbial decomposition of organic material. Phosphate, the major limiting nutrient in freshwater ecosystems, may exceed 1 mg dm^{-3}. This combination of increased nutrient and light availability stimulates the growth of algae, particularly the filamentous *Cladophora*. Large beds of this algae may come to dominate the recovery zone. Algal photosynthesis further replenishes the supply of dissolved oxygen. However, respiration of large amounts of algae will deplete free dissolved oxygen in the water at night. Consequently, this zone is characterised by large diurnal fluctuations in the concentration of dissolved oxygen.

The *Cladophora* beds support large numbers of animals including midge (chironomid) larvae (Fig 13.8(a)) and, if the bottom is muddy, *Tubifex* worms (Fig 13.8(b)). These organisms feed on the abundant supplies of bacteria and organic material present both in the water and the sediment. This further accelerates the recovery process by reducing the amount of

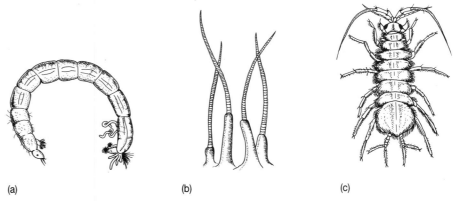

(a)   (b)   (c)

Fig 13.8 **(a)** Chironomid larva. The colour depends on the blood pigments. The blood larva is red because it contains haemoglobin. Others are grey and green. They swim by looping and unlooping.
(b) Tubificid worms are annelids that live inside cases. Their blood is red because it contains haemoglobin.
(c) The water louse, *Asellus*, is grey-brown in colour and has a flattened body. It feeds on dead and decaying material.

ORGANIC POLLUTION

suspended organic material. As levels of dissolved oxygen increase further the water hog louse, *Asellus aquaticus* (Fig 13.8(c)), becomes a dominant member of the invertebrate fauna, and coarse fish, like bream, start to reappear.

Clean water zone

Finally, the entire clean water fauna and flora, including game fish like salmon, return. This may not happen for a considerable distance below the point of discharge. For example, the complete recovery of the algal community below sewage discharges on two English rivers, the Tame and Trent, took 70 km and 56 km respectively.

ANALYSIS

Pollution in the River Trent

This exercise requires you to interpret graphs using information you have learnt so far in this chapter.

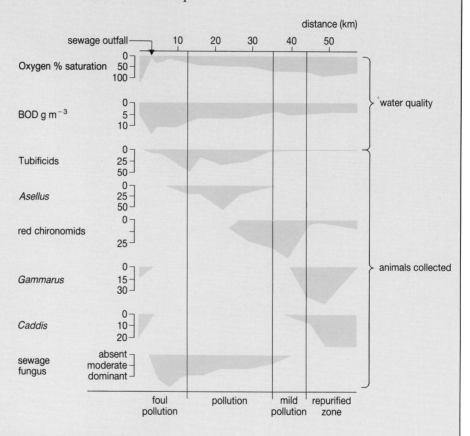

Fig 13.9 Pollution in the River Trent.

Carefully examine Fig 13.9 and answer the following questions.
(a) What indicates that the sewage contained a large amount of organic matter?
(b) Which animals appear to be the most tolerant of pollution?
(c) What is the most likely explanation for the changes in the distribution of organisms downstream of the sewage discharge?
(d) What characterises the habitat of sewage fungus?
(e) Why do chironomids and tubificids do so well in water polluted with organic material?

Recovery

The exact pattern of recovery will of course vary from river to river. Indeed, some rivers may never fully recover as they receive further loads of BOPs, so reducing the level of dissolved oxygen again. However, the encouraging picture which emerges from Fig 13.4 is that rivers do have the capacity to cleanse themselves; and that this capacity is based on natural processes: dilution, sedimentation and biodegradation. This raises the interesting possibility that if we could engineer such processes in a treatment works so that an effluent had already undergone them before it was discharged then the problems we have witnessed would never arise. This is the basis of sewage treatment which is dealt with in another book in this series by Jane Taylor, *Micro-Organisms and Biotechnology*. Furthermore, the characteristic changes in the fauna and flora associated with pollution by oxygen-demanding wastes suggests the possibility of biological monitoring of water quality. This is discussed further in Chapter 15.

QUESTIONS

13.6 Account for the changes in BOD, NH_4 and NO_3 levels shown in Fig 13.4.

13.7 Domestic sewage has a BOD of 250–350 g m^{-3}. However, the BOD of the water receiving this sewage is unlikely to have a BOD greater than 10 g m^{-3}. What causes this reduction in BOD?

13.8 (a) What is 'biochemical oxygen demand'? Describe how it may be measured and give the units.

Table 13.3

Distance from outfall (arbitrary units)		Chemical variable (values shown in arbitrary units)	
		A	B
upstream	2	100	10
	1	100	10
outfall	0	98	80
	1	30	77
downstream	2	27	67
	3	38	50
	4	58	34
	5	72	20
	6	88	17

(b) The figures displayed in Table 13.3 represent two chemical variables, A and B, measured above and below a sewage outfall into a river.
(i) Plot these figures on a graph, using the upper half of the graph paper.
(ii) Suggest the identity of variables A and B and give your reasons.
(iii) On the lower half of the graph paper, draw three lines to represent other variables (one chemical, one physical and one biological). Explain the shape of each curve.

(c) State two factors which may affect the recovery of a river from organic pollution downstream of a discharge point.

SUMMARY ASSIGNMENT

1. Keep a record of the sources of biodegradable organic pollutants.

2. Produce a chart or diagram to summarise the changes which occur in the flora and fauna below a sewage discharge. Incorporate the information in Fig 13.4 into your diagram. Keep a record of your answers to Question 13.6 and Analysis: Pollution in the River Trent to remind you of the factors which cause the changes shown in your diagram.

3. Using your library and the information given in this chapter explain why animals like chironomids and hoverfly larvae survive in water with a low oxygen concentration while other animals like mayflies and stoneflies cannot.

4. Explain how dilution, sedimentation and biodegradation produce the 'self cleansing' process seen in Fig 13.4.

Chapter **14**

ACID RAIN, EUTROPHICATION AND TOXIC POLLUTION

Human activities, industry and agriculture all release chemicals into the environment. The biological effects of these chemicals, the pathways they follow and the repercussions for people and wildlife are difficult to predict. Who could have predicted that irrigating crops with water from an old mine spoil heap would have caused the painful symptoms of Itai-Itai (literally ouch-ouch) disease, the result of cadmium poisoning, in over 1000 Japanese women. The industrial revolution has brought many countries great wealth but it has also left a legacy of acidified lakes. Modern agricultural techniques produce more food more cheaply than ever before yet the same fertilisers needed to boost production can completely deoxygenate lakes. Every single one of us carries a few micrograms of DDT around in our body, and newly fallen Arctic snow is contaminated with polychlorinated biphenyls (PCBs). How have we reached this state of affairs? What can be done about it? How much will cleaning up cost? These are some of the questions addressed in this chapter.

LEARNING OBJECTIVES

After completing the work in this chapter you will be able to:

1. define and explain the following terms: acid rain, eutrophication, toxicity, (LC_{50}, LD_{50}), acute toxicity, chronic toxicity; sublethal effects and synergism;

2. account for, and explain the consequences of, the acidification of freshwater ecosystems;

3. explain the origins and possible cures for cultural eutrophication;

4. discuss the concept of toxicity and explain how some toxic compounds exert their effects.

14.1 ACID RAIN

Gases, normally present in the atmosphere, dissolve in water vapour to give acid solutions and so unpolluted rain water is naturally acidic with a pH of 5.2–5.6. For example:

Carbon dioxide produces a weak solution of carbonic acid:

$$CO_2(g) + H_2O(l) = HCO_3^-(aq) + H^+(aq)$$

Sulphur dioxide from volcanic eruptions and trimethyl sulphides produced by marine phytoplankton produce sulphuric acid:

$$2SO_2(g) + O_2(g) + 2H_2O(l) = 2H_2SO_4(aq) = 4H^+(aq) + 2SO_4^{2-}(aq)$$

Nitrous oxides, produced during electrical storms, give nitric acid:

$$4NO(g) + 3O_2(g) + 2H_2O(l) = 4HNO_3$$

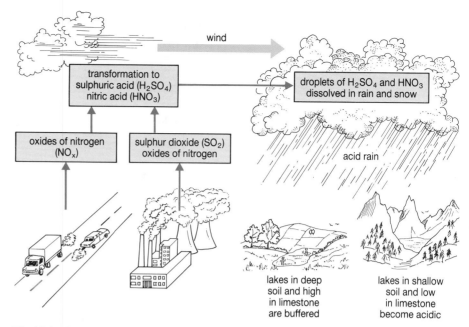

Fig 14.1 The origins of acid deposition. Note that potentially acidic materials can be deposited in either a dry form, dry deposition, or in solution, acid rain.

It follows therefore that the acidity of rain will increase if concentrations of sulphur dioxide, produced by, for example, power stations, and nitrous oxides, produced by high temperatures in car engines, also increase (Fig 14.1). Since 1950, sulphur dioxide and nitrous oxide emissions in Europe have doubled leading to the formation of rain with a pH < 5 (Fig 14.2). On 10 April 1974, a rainstorm at Pitlochry in Scotland had a pH of 2.4–an acidity greater than that of vinegar!

Acid rain represents a major political problem, as prevailing winds carry pollution across international borders. For example, only 8% of the sulphate falling on Norway is due to the activities of Norwegians – 17% comes from the UK. By contrast, 79% of UK sulphate deposition is of British origin – none comes from Norway.

ANALYSIS

The origin of acid rain

This is a comprehension exercise involving the analysis of tabulated data.

Table 14.1 presents data on the composition of rain water from central Scandinavia during the 1950s and 1970s. How does this data corroborate our explanation of the origin of acid rain?

Table 14.1 Ion concentrations and pH of rain water from Scandinavian and US sites in 1956 and 1974.

Site	pH	$[H^+(aq)]$ $10^{-6}\,mol^{-1}$	$[SO_4^{2-}(aq)]$ $10^{-6}\,mol^{-1}$	$[NO_3^-(aq)]$ $10^{-6}\,mol^{-1}$	$[HCO_3^-(aq)]$ $10^{-6}\,mol^{-1}$
inland Scandinavia (1956)	5.4	4	15	0	6
inland Scandinavia (1974)	4.3	48	26	26	0
inland northeastern United States (1974)	3.9	114	55	50	0

The extent of the problem

Rivers in Norway, Sweden, southern Scotland and central Wales have all been reported to have a pH < 5.0. Lakes affected by acidification occur in Ontario, New England, the Galloway region of Scotland and the Brecon Beacons of Wales. Scandinavia is particularly badly affected:

- Norway – lakes in an area of 13 000 km² are totally devoid of fish.
- Sweden – 18 000 lakes have a pH < 5.5 for part of the year; fish stocks have been reduced or eliminated in half of them.

The acidification has been rapid. Analysis of diatoms in Scottish lake sediments indicates that the acidification started with the industrial revolution. In Norway, rivers with a pH of 5.0–6.5 in 1940 had a pH from 4.6 to 5.0 by 1976, with a pH as low as 3.0 during dry spells or after snow melts in spring. Remember, pH is measured on a logarithmic scale, so a change in pH from 6.0 to 5.0 represents a ten-fold increase in hydrogen ion concentration, a change from 6.0 to 4.0 a hundred-fold increase and so on.

But why does acidification only occur in these areas? Why do lakes in other areas, like southeastern England, where the rain has an equally low pH (see Fig 14.2), not become acidified? The answer lies in the soil. Acidified lakes and streams are found in areas where the rocks are hard and poorly weathered. The resulting thin, acid soils contain few bases, like calcium carbonate, which could neutralise the acid rain. Acidification is only a problem when the soils are themselves acid.

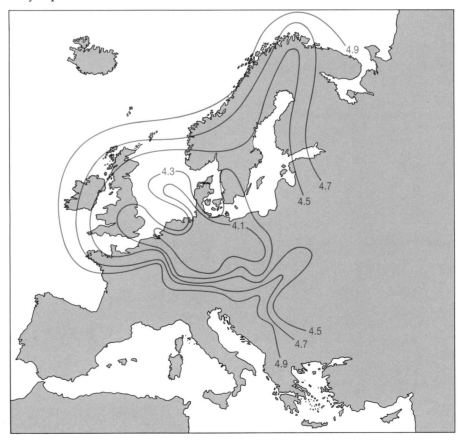

Fig 14.2 Mean pH values for rainfall over western Europe in the period 1978–82.

The effects of reduced pH

These are summarised in Table 14.2. Of particular importance is the loss of fish stocks, especially salmon and trout, in acidified waters. This is partly the result of the fish dying and partly due to a failure of reproduction. For

ACID RAIN, EUTROPHICATION AND TOXIC POLLUTION

Table 14.2 Damage to the aquatic environment which occurs with decreasing pH.

Range	Effects
in the range pH 8.0–6.0 a decrease of:	
< 0.5 pH units	very slight change in biotic environment
0.5–1.0 pH units	detectable alterations in community composition, some species eliminated
in the range pH 6.0–6.5	reduction in species numbers, and among remaining species considerable alteration in ability to withstand stress
in the range pH 5.5–5.0	many species eliminated and a few pH-tolerant invertebrates will become abundant
pH < 5.0	decomposition of organic detritus is severely impaired, most fish species eliminated
pH < 4.5	all above changes greatly exacerbated, fish cannot survive

example, salmon eggs fail to hatch if exposed to pH 4.0–5.5 at the stage where the embryo has developed eye pigment. This is probably due to the inhibition of the enzyme chorionase (optimum pH 8.5) which dissolves the egg membrane, allowing the larval fish to escape.

Low pH also:

- interferes with ion regulation, for example increasing Na^+ loss through fish gills;
- reduces the efficiency with which haemoglobin takes up oxygen;
- increases mucus deposition on the gills, so increasing the distance over which oxygen must diffuse from the water to the gill surface.

The importance of aluminium and calcium

Research has shown that more fish die in acidic waters when the levels of calcium are low and, especially, if aluminium levels are high. As acid rain percolates through soil the hydrogen ions (H^+) it contains interact with clay minerals, displacing aluminium ions (Al^{3+}) which leach out into rivers and lakes. Aluminium has a number of effects, including:

- raising the pH thresholds at which the physiological effects discussed above occur;
- killing fish – for example, an aluminium concentration of 200 µg dm^{-3} at pH 5 appears to be lethal to brown trout;
- precipitating phosphate from soil water. The phosphate then remains in the soil rather than entering the lake. Since phosphate is an essential plant nutrient, phytoplankton can no longer grow, so the **oligotrophic** (nutrient-poor) lake water becomes clear since such water bodies have a low biomass of organisms in them.
- flocculating particles, including phytoplankton, which then fall to the bottom of the lake.

Low pH also interferes with the uptake of calcium. This means that crustaceans like crayfish, *Gammarus* and *Asellus*, which require calcium to produce their exoskeleton, are absent from acid waters. Some evidence suggests that birds, like pied flycatchers, which feed on calcium-deficient, aluminium-rich aquatic inserts may lay thin-shelled eggs which reduces hatching success.

Generally, then, the effects of acidification are to:

- reduce the species diversity of the freshwater communities;
- reduce fish stocks;
- increase the organic material present in the sediments as detritivores, like *Asellus* and snails, disappear;
- decrease the amount of plant nutrients in the water and so reduce phytoplankton growth.

These effects are mainly the result of the effects of acid rain on the soil, particularly the release of aluminium, and are not simply produced by lowering the pH of the water.

Solving the problem

The only real solution is to reduce the emission of the gases which cause acid rain. The speed with which this can be done depends on how much governments and, ultimately, you and me are willing to spend. In 1986 the UK government announced a projected expenditure of £600 million on chemical plant to reduce sulphur dioxide in emissions from power stations. The Norwegian government claimed that this was far too little too late. Now the newly privatised power industry seems to be moving away from this commitment and towards a strategy of burning gas and using low sulphur coal as a means of reducing sulphur dioxide emissions.

Fig 14.3 Limestone pavement in the Yorkshire Dales. Removing this readily available source of limestone would damage a beautiful area. Is this a price worth paying for curing acid rain?

We may also have to pay another price for cleaning up acid rain. Sulphur dioxide scrubbers need calcium carbonate (limestone) to work. This rock must be quarried, for example, from the Yorkshire Dales (Fig 14.3). So to cure acid rain may require many new, unsightly quarries to be created, as well as new roads to take large lorries carrying the limestone, in some of Britain's most beautiful landscape.

Even if we were to stop emissions of sulphur dioxide and nitrous oxide now, many catchments, acidified over many years, have such a low buffering capacity that high H^+ and aluminium run-off will continue for many years. Liming these catchments and the acidified water may help but it is very expensive. Nonetheless, liming has succeeded in reducing the acidity in at least one Scottish lake to a level where trout can survive and breed.

ACID RAIN, EUTROPHICATION AND TOXIC POLLUTION

14.1 Extremes of pH, or rapid changes in pH, are detrimental to most aquatic organisms. Common aquatic plant and animal populations flourish within a fairly narrow range of pH values, although most will tolerate, at least for a while, variations between pH 6 and 9. Explain these observations.

14.2 Excess nitrates from fertiliser use and sewage seem to be stimulating the growth of phytoplankton in the North Sea. Explain how this could be responsible for at least some of the acidification of Swedish lakes.

14.3 The hatching and juvenile stages of trout and salmon coincide with periods of snow melt in the spring. Why will this have a particularly devastating effect on recruitment in these species?

14.4 Fish exposed to acidic conditions may die of oxygen starvation. Why?

14.5 Explain the following observations.
 (a) Acidified lakes often become oligotrophic (nutrient-poor).
 (b) Peat deposits (undecomposed plant remains) build up in acid lakes.
 (c) Acidification is not a problem in chalky and limestone areas.

14.2 EUTROPHICATION

To renew all the water in a lake or reservoir may take up to 100 years compared to a few days in a river. Consequently, lakes are particularly susceptible to pollutants. The most widely reported problem is accelerated eutrophication.

The cause

Eutrophication is a natural process. Over time, lakes gradually become enriched in plant nutrients, e.g. phosphates and nitrates, as a result of natural rock erosion and run-off from the surrounding catchment, leading to an increase in plant biomass. However, human activities (Fig 14.4) can greatly increase the rate of this process from a few thousand years to a few

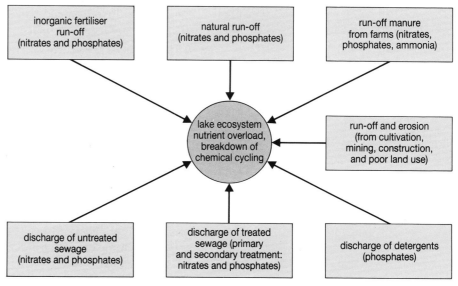

Fig 14.4 The causes of cultural eutrophication.

ACID RAIN, EUTROPHICATION AND TOXIC POLLUTION

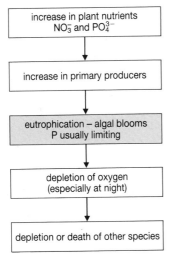

increase in plant nutrients
NO_3^- and PO_4^{3-}

↓

increase in primary producers

↓

eutrophication – algal blooms
P usually limiting

↓

depletion of oxygen
(especially at night)

↓

depletion or death of other species

Fig 14.5 Stages in the process of eutrophication. Note the increase in both nitrate and, more importantly, phosphate concentrations in the water body.

Fig 14.6 A thick crust of algae growing in the severely eutrophicated water of an inland lake.

decades. Phosphate is the major limiting nutrient in freshwater so eutrophication is often associated with this particular plant nutrient. The process is shown in outline in Fig 14.5.

Table 14.3   The effects of eutrophication on the receiving ecosystem and the human problems associated with these effects.

Effects
1. Species diversity decreases and the dominant biota change
2. Plant and animal biomass increases
3. Turbidity increases
4. Rate of sedimentation increases, shortening the life-span of the lake
5. Anoxic conditions may develop

Problems
1. Treatment of drinking water may be difficult and the supply may have an unacceptable taste or odour
2. The water may be injurious to health
3. The amenity value of the water may decrease
4. Increased vegetation may impede water flow and navigation
5. Commercially important species (such as salmonids and coregonids) may disappear

The effects

The changes associated with this cultural eutrophication are summarised in Table 14.3. The associated problems include the following.

- Increased growth of rooted plants, e.g. reeds, so that river channels and lakes become overgrown.

- Algal blooms (Fig 14.6). Explosive growth of algae and cyanobacteria ('blue-green algae') during the summer produce water the colour of pea soup. Furthermore, some of these organisms, e.g. *Prymnesium parvum*, release substances which may be poisonous (populations of *P. parvum* in excess of 10^4–10^5 cells per cm^3 are sufficient to kill fish) and make the water taste and smell awful. This alga has caused severe problems in fish ponds in Israel and has killed fish in Hickling Broad in East Anglia.

- The growth of phytoplankton, at high concentrations of plant nutrient, may shade out macrophytes (large plants) which grow on the bottom of the lake leading to a reduction in habitat for aquatic invertebrates (e.g. dragonflies), birds and fish.

- A depletion of dissolved oxygen in the water when the algae die, fall to the bottom of the lake and are decomposed by aerobic bacteria.

- The disappearance of fish, like trout, and invertebrates which are sensitive to low levels of dissolved oxygen, while species which can tolerate low oxygen levels, like carp, become predominant.

- The hypolimnion (bottom water) will become totally anoxic, if sufficiently high levels of nutrients are present, as a result of the decomposition of large amounts of dead algae. Under these conditions, anaerobic bacteria predominate, producing foul smelling gases like hydrogen sulphide.

- Blockage of filters in water treatment works by algal cells, while the smallest algal cells and cyanobacteria may pass through the filters and enter the domestic drinking water supply. Such cells, decomposing in the pipes, impart tastes and smells to the water. Bacteria, fungi and invertebrates, feeding on the decaying algal cells and each other, may block pipes.

ACID RAIN, EUTROPHICATION AND TOXIC POLLUTION

The nitrate controversy

High levels of nitrate derived from agricultural run-off (see Chapter 9) in drinking water may be a potential health risk. For example, children under six months of age may develop methaeomoglobinaemia if their bottle-fed milk is made with water high in nitrate. The low pH of the gastric juices of such children favours the conversion of nitrate ions to nitrite ions which are readily absorbed into the blood. Here the nitrite reacts with the iron in haemoglobin producing methaemoglobin, so lowering the oxygen-carrying capacity of the blood. Above 25% methaemoglobin there is blueing of the skin (cyanosis). Levels of 60–85% methaemoglobin result in death.

Levels of nitrate in excess of 100 mg NO_3 m^{-3} in water may result in the formation of nitrosamines in the acidic conditions found in the human stomach. Nitrosamines are probably carcinogens (cancer-causing agents) but the epidemiological data is lacking to substantiate the risk associated with these compounds.

It must be stressed that even in high nitrate areas, e.g. East Anglia, methaeomoglobinaemia is very rare and epidemiological evidence suggests little effect, if any, on gastric cancer in the UK. Recent data suggests that water nitrate levels can only affect the incidence of these diseases when there is malnutrition, in particular low vitamin C levels. Nonetheless, we should not be complacent. The European Health Standard for nitrates in drinking water is less than 45 g m^{-3}. Water taken from rivers in lowland England frequently has nitrate levels in excess of 100 mg dm^{-3} and has to be diluted with water low in nitrates before it can be used.

Cleaning up

The effects of cultural eutrophication can be controlled by **input methods**, to reduce the input of nutrients, particularly phosphorous, and **output methods**, to clean up water bodies already suffering from excessive eutrophication.

Input control methods include the following.

- Tertiary treatment of sewage which can remove 90% of phosphates in sewage effluents. However, this is very expensive.

- Banning or reducing the use of polyphosphates, e.g. sodium tripolyphosphate, as builders in washing powders and other household cleaning agents. This one source contributes some 30–50% of the phosphate in sewage effluents.

- Control the use and run-off of fertilisers from farm land by reducing the use of nitrate fertilisers.

- Direct waste water away from lakes to streams and the seas. This may of course simply transfer the problem from one place to another. For example, waste water entering Lake Washington near Seattle was diverted into the nearby Puget Sound. Lake Washington recovered. Puget Sound now suffers from eutrophication.

Output control methods are applied to lakes already eutrophicated and include the following.

- Precipitating phosphates from the water with aluminium or iron salts. This treatment will also flocculate particular matter, so clearing the water. The treatment has to continue as long as phosphate continues to enter the water. Aluminium sulphate is used to precipitate organic material in water treatment works. In 1988, 20 tons of this substance was put into the wrong tank at the Camelford treatment works in Cornwall. As a result the aluminium salt entered the drinking water supply causing one of the largest mass poisonings in the UK's history.

Fig 14.7 A portable aerator being used to replenish oxygen in a fish pond. Why would this not be an appropriate method for aerating the hypolimnion of a eutrophicated lake?

- Aerate lakes and reservoirs to prevent oxygen depletion (Fig 14.7). This involves bubbling air into the hypolimnion (i.e. the bottom water of the lake) without mixing the nutrient-rich bottom water with the sunlit surface waters. Although this increases the amount of oxygen available to fish it is expensive and only treats the symptoms, since it does not remove excess plant nutrients.

- Remove plant material, which has accumulated phosphorous during the growing season, before it decomposes, so releasing the phosphorous back to the water. This is an expensive method which will take many years to have an effect. It also presents a new problem: what to do with the vast amount of phosphate-rich plant material.

These output control methods have had little success since most of the phosphorous introduced into a lake ends up in the bottom sediments. So phosphorous removed from the water is replaced by that from the sediments. To clean up you have to get rid of this sediment.

A success story

The Swedish Lake Trummen had become highly eutrophicated after receiving sewage and industrial effluents for over 30 years. Every year, 8 mm of new, black, sulphurous mud accumulated on the bottom. In the winter the lake froze and its water, only 2 m deep, became deoxygenated, leading to fish kills. Some sewage and industrial effluents were diverted in 1958 but by 1968 the lake still produced an enormous amount of algae in the summer, suffered deoxygenation and fish kills in the winter, and was highly turbid. The growth of algae was supported by phosphate released from the sediments. This internal source produced 177 kg P per year compared with 3 kg per year from external sources.

In 1970 and 1972 a slurry of surface sediments and water was pumped from the lake into specially constructed lagoons. Here the sediment, plus its phosphorous, settled out and was later disposed of as fertiliser. The water, after treatment with aluminium salts, was returned to the lake. The results were dramatic. The water cleared and winter deoxygenation no longer occurred.

Such dredging can only be applied to shallow lakes and has the disadvantage of resuspending toxic pollutants like methyl mercury. Nonetheless, it does suggest a way forward, providing there is the political will and the money.

ACID RAIN, EUTROPHICATION AND TOXIC POLLUTION

Eutrophication

This exercise will give you practice in interpreting tabulated data.

Table 14.4 compares certain features of an oligotrophic and an eutrophic lake.

(a) Define 'oligotrophic' and 'eutrophic' and explain how the conditions may arise.

(b) Describe briefly a simple method for measuring the penetration of light.

(c) Explain the relationship between the figures for light penetration and chlorophyll.

(d) Explain the significance of the figures for rate of depletion of hypolimnetic oxygen and suggest what the units could be.

(e) Why were phosphorus values measured in winter and other values measured in summer?

(f) What other inorganic variable indicates whether a lake is oligotrophic or eutrophic?

(g) Give two other features of an eutrophic lake.

Table 14.4

Feature	Period when measured	Oligotrophic lake	Eutrophic lake
phosphorus ($\mu g\ dm^{-3}$)	winter	10–15	20–30
chlorophyll ($\mu g\ dm^{-3}$)	summer	2–4	6–10
light penetration (m)	summer	3–5	1.5–2
rate of depletion of hypolimnetic oxygen (arbitrary units)	summer	250–310	330–400

14.6 The level of eutrophication that is socially acceptable is partly a function of cultural and economic needs. Explain why someone farming carp would accept, indeed engineer, a higher level of eutrophication than someone who wanted to use a lake for swimming.

14.7 Why is it important that during aeration of eutrophicated water the hypolimnion does not mix with the epilimnion?

14.8 (a) Why may algal blooms cause large daily fluctuations in the dissolved oxygen content of a lake's surface waters?

(b) What effect will such fluctuations have on fish living in the lake?

(c) Why did Lake Trummen suffer fish kills during the winter and not the summer?

(d) Fig 14.8 compares the ecological consequences of organic pollution and eutrophication.
   (i) Account for the differences shown in the figure.
   (ii) Why is the end result, i.e. oxygen depletion and fish kills, the same?

14.3 TOXIC POLLUTION

Some commonly used terms applied to toxic pollution are given in Table 14.5

The toxicity of all materials depends on their concentration. It is therefore incorrect to classify substances as toxic or non-toxic without referring to their concentration in the environment. Indeed, some essential components

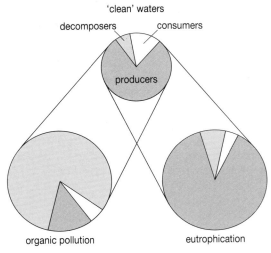

'clean' waters

decomposers   consumers

producers

organic pollution   eutrophication

Fig 14.8 A comparison of the ecological effects of organic and inorganic pollution.

Table 14.5   Some commonly used terms applied to toxic pollution.

Term	Definition
lethal	death by direct poisoning;
sublethal	death caused indirectly by affecting growth, reproduction or an activity leading to a reduction in the number of species and/or individuals;
acute	causing an effect (often death) within a short period;
chronic	causing an effect (lethal or sublethal) over a prolonged period of time;
accumulative	effect increased by successive doses.

of an animal's diet, e.g. zinc, may become toxic at high concentrations. Toxicity will also be affected by:

- the biodegradability of the substance;
- the actual exposure (dose × time exposed to dose) suffered by an organism;
- the bioavailability of the substance – for example, heavy metals may be bound to the substrate and so be unavailable for uptake by animals or plants;
- the rate of uptake and elimination of a substance from an organism, which determines the retention time;
- the amount of biomagnification along a food chain.

Measuring toxicity

This is usually achieved by means of LC_{50} or LD_{50} tests. LC_{50} is the concentration of a pollutant which kills 50% of a test population in a given time period. LD_{50} is the dose which kills 50% of a test population in a given time period. Fig. 14.9 shows some imaginary data for a pesticide. The 22- week LC_{50} for this compound would be about 80 g m^{-3}. Some real LC_{50} and LD_{50} values are given in Table 14.6.

The application of both of these laboratory-derived measures to field populations is increasingly being called into question but they do give us some standards (Table 14.7). Particular limitations include the following:

- The final effects of a toxic substance depend on environmental conditions, like pH, temperature and water hardness. For example,

ACID RAIN, EUTROPHICATION AND TOXIC POLLUTION

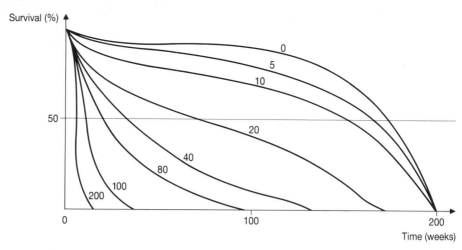

Fig 14.9 Graphs of percentage survival against time for toxicity tests on various concentrations (g m^{-3}) of a pesticide. The point of intersection of the horizontal red line with the curves represents lethal concentrations for 50% of the population (LC$_{50}$).

Table 14.6

	LD$_{50}$	Examples
highly toxic	1–10 mg kg^{-1} body weight	arsenic
moderately toxic	10–100 mg kg^{-1} body weight	cadmium, copper, lead, mercury
slightly toxic	100–1000 mg kg^{-1} body weight	aluminium, molybdenum, zinc
relatively harmless	1000 mg kg^{-1} body weight	sodium, iodine, calcium, potassium

Table 14.7 Concentrations of various insecticides lethal to 50% of fish exposed to them for 96 hours.

Insecticides	LC$_{50}$
Aldrin	0.033
DDT	0.032
Dieldrin	0.016
Malathion	12.2
Parathion	1.4–2.7

heavy metals are generally less toxic in calcium-rich, hard water; a nickel–cyanide complex is 500 times more toxic to fish at pH 7.0 than at 8.0; ammonia is ten times more toxic at pH 8.0 than 7.0.

- Two pollutants, both present at safe limits according to LC$_{50}$ tests, may have an additive or **synergistic** effect. For example, an equal parts solution of cadmium and zinc is twice as toxic as either metal on its own.

Major types of toxic pollutant

Substances generally considered to pose environmental hazards include heavy metals, man-made organic and inorganic chemicals, fibres, particulate material and oils. Two particular problems are associated with these materials.

1. Synthetic organic substances, e.g. DDT and PCBs, unknown in nature, tend to be non-biodegradable. As a result these toxic, often lipid-soluble substances can accumulate in animals and plants.
2. Human-induced rates of environmental contamination by naturally occurring substances, e.g. heavy metals, are much greater than natural rates.

Toxic pollutants exert their biological effect in a number of ways. The most common mode of action is enzyme inhibition. The mode of action of pesticides is discussed in Chapter 10. Heavy metals, such as lead and mercury, affect the central nervous system although their exact mode of action is unknown.

Case Study: Polychlorinated biphenyls

PCBs constitute a group of some 70 oily, synthetic compounds which are stable at temperatures up to 800°C, resistant to acids, bases and oxidants, and only sparingly soluble in water. They were widely used as heat transfer fluids, insulators, hydraulic fluids and plasticisers, and were incorporated into a range of products from paints to fluorescent lights.

The toxicity of these apparently useful substances was first noticed in 1968 when 1300 Japanese developed skin lesions (chloracne) and suffered liver and kidney damage after they had eaten rice oil contaminated with PCBs that had leaked from a heat exchanger. In 1973, 225–450 kg of the related polybrominated biphenyls was accidentally added to animal feedstuffs in the United States. The resulting contamination of livestock necessitated the destruction of 289 800 cattle, 5920 pigs, 1470 sheep and 1.5 million chickens.

Insoluble in water but soluble in fats, PCBs may biomagnify in aquatic food webs. They enter the body through the skin, food and inhalation. The long-term effects of PCBs are unknown but in laboratory animals they are known to cause liver damage, kidney damage, gastric disorders, reproductive disorders, skin lesions and tumours.

The long-term problems presented by PCBs are enormous. While their production was voluntarily abandoned in the United States in 1974, an estimated 68 000 tonnes of PCBs had already been released into the environment in the United States alone. Consequently, practically every environmental sample, from water to Arctic snow, contains traces of PCBs. They are still to be found in closed electrical equipment like electrical transformers. When such transformers are involved in fires the PCBs can be released into the environment. There is an estimated 341 million kilograms of PCBs in transformers in the United States.

Disposal of such inert materials is difficult. They have to be incinerated at high temperatures in carefully controlled furnaces (Fig 14.10). The UK leads the world in such technology. Consequently, cargoes of PCBs are often sent here to be incinerated. This led to a storm of protest in 1989 which succeeded in turning back cargoes of PCBs coming from Canada.

The PCB story highlights a difficult problem. The chemical industry continually produces new substances which are of great, potential benefit to people. Unfortunately, we cannot predict the ecological effects of these substances before they are released into the environment. More research is desperately needed to enable us to predict the effects of such material on the environment before they are manufactured in large quantities. Modern legislation, by imposing stricter testing procedures, is helping to prevent another PCB story.

Fig 14.10 Some dangerous chemicals, like PCBs, can be rendered totally harmless provided they are burnt under controlled conditions in an incinerator like this one in Wales. However, incineration under the wrong conditions can lead to dangerous chemicals like dioxins being emitted from the stack.

ACID RAIN, EUTROPHICATION AND TOXIC POLLUTION

14.9 You have been asked to decide whether an electrical factory should be allowed to continue discharging an effluent which contains a number of heavy metals, including mercury, zinc and cadmium, into a local, acidified stream. The factory, located in an upland area of Britain, employs 2000 people and is the only major source of work in the area. You have the results of LC_{50} tests on individual heavy metals conducted in a laboratory using trout kept in water from a lake fed by a chalk stream.

(a) How useful would this toxicity data be in helping you to make your decision?

(b) What further data would you try to collect before making your decision?

(c) Putting in expensive plant to totally eliminate metals in the effluent would force the factory into bankruptcy and closure. How might this influence your decision?

SUMMARY ASSIGNMENT

1. Ensure that you have definitions, explanations and appropriate examples of the terms listed in the learning objectives.

2. Prepare a report for a group of MPs explaining the benefits and costs of fitting sulphur scrubbers to power stations. Remember that you are not dealing with scientists so you will need to explain why apparently small changes in pH have such a devastating effect. How would you respond to your report if you represented:
   (a) a constituency in southern England and your constituents were complaining about the level of their taxes·
   (b) a Scottish upland constituency dependent on a tourist industry based on salmon fishing;
   (c) a constituency in the Yorkshire Dales where many local people were leaving because there were no jobs?

3. Compare and contrast the processes which result in deoxygenation when raw sewage is occasionally discharged into a river and the run-off of fertilisers from agricultural land continually enters a lake.

4. The list of toxic pollutants is enormous. Select two organic and two inorganic substances from the lists given in this chapter. Compile a dossier of information on each substance. Include information on chemical composition, biodegradability, toxicity sources, effects, concentration in the environment, accumulation in organisms and food chains, pollution incidents involving your chosen substance, disposal/clean-up methods. If you split your class into research groups you will be able to compile a lot of information. Produce a display to illustrate your findings.

Chapter 15

MONITORING WATER QUALITY

The easiest and cheapest way for an unscrupulous company to get rid of its waste is to dump it (Fig 15.1). Unfortunately, too much untreated effluent is still disposed of in this way. However, it would be unrealistic to think that we need to treat all industrial effluents so that they have no potential pollutants in them. What needs to happen is for the effluents to be treated so that they have an acceptable level of pollutants in them, i.e. levels which will not cause unacceptable biological damage. This of course raises the question of what is acceptable. In Britain this is usually worked out between the industry concerned, the local water company and the National Rivers Authority (NRA) using what is called a **consent**. This sets the maximum levels of pollutants which the company is allowed to dispose of into a river or stream. The National Rivers Authority is then primarily responsible for monitoring water quality and ensuring that the company stays within its consent limits. This chapter, then, is concerned with how the chemists and biologists who work for the NRA, water companies and industry monitor pollutant levels to ensure companies stay within their consent limits and that the water in our rivers remains, relatively at least, safe.

Fig 15.1 This effluent, which probably contains a complex cocktail of potentially toxic chemicals, needs to be carefully monitored. Companies are only allowed to put waste, with a known and regulated composition, into streams with the explicit permission of the National Rivers Authority. However, 'accidents' happen!!

LEARNING OBJECTIVES

After completing the work in this chapter you will be able to:

1. appreciate the need to monitor pollution levels;

2. outline some chemical and physical tests used to monitor water quality, including BOD, COD and permanganate value;

3. discuss the advantages and disadvantages of biological monitoring;

4. calculate and interpret biotic and diversity indices.

15.1 ROUTINE TESTING OF WATER

The quality of untreated and treated water, sewage and industrial effluents is assessed using a battery of physical, chemical and biological tests. Some of these are summarised in Table 15.1. Pollution monitoring often involves measuring incredibly low concentrations of substances; less than 1 ppb is common. To achieve this, sophisticated instruments are needed. Some of these are shown in Fig 15.2. While a good analyst, with a well equipped laboratory, can identify an enormous number of pollutants in water usually this is not necessary. Time and money can be saved by carrying out a limited range of tests which experience tells us produces useful results. The precise range of tests will depend on the source of the sample and the purpose of the analysis. For example, textile processing needs water free of iron and manganese, so water used in this industry must be analysed for these elements.

Table 15.1   Some important water quality tests, their chief significance and general means of measurement.

Quality parameter	Significance	General method of analysis: expression of result
colour (apparent)	suspended and dissolved solids; organic matter	comparison with platinum–cobalt standard: Hazen colour units
odour	a large amount of compounds, e.g. ammonia, chlorine	subjective perceived odour: threshold number
turbidity	estimate of suspended matter	comparometric: a standard unit called the Formazin turbidity unit
dissolved oxygen	approximate extent of easily oxidised organic matter	titration or electrode: per cent saturation or $g\ m^{-3}$
suspended and settleable solids	turbidity; treatment efficiency	gravimetric: $g\ m^{-3}$ or cm^{-3} or dm^{-3}
pH value	intensity of acid or alkali present; strength of effluents affects many chemical and biological properties	titration; electrode; colorimetric; pH values 0–14
biochemical oxygen demand	extent of biodegradable organic matter	measurement of dissolved oxygen before and after incubation for 5 days at 20°C; grams of oxygen consumed per m^3
permanganate values 3-minute 4-hour	oxidisable inorganic matter and very easily oxidised organic matter	titration: grams of oxygen consumed from a standard permanganate solution per m^3
chemical oxygen demand	estimate of organic matter present (more accurate than permanganate value)	titration: grams of oxygen consumed from standard dichromate solution per m^3

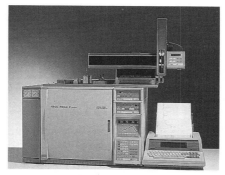

Fig 15.2   (a) A gas chromatograph used for testing water to detect organic compounds like pesticides. The sample is injected into the top of the machine and the concentrations of chemicals in the sample are displayed on the chart recorder.
(b) Atomic absorption spectrophotometers are used to determine concentrations of heavy metals in water samples. Many samples are placed in individual glass beakers on the tray at the front of the machine and the machine will then automatically process them one by one. Results are recorded on the chart recorder on the right.

Chemical and physical tests

Routine monitoring of river water includes measurement of temperature at the time of sampling, pH, dissolved oxygen, amount of suspended and total solids and alkalinity (the actual amount of alkali, usually bicarbonate, in solution). If you suspected pollution by a sewage effluent then you might also include tests for ammonia, organic nitrogen, nitrite and nitrates, chloride, phosphorous, detergents and organic matter. Inclusion of chloride in this list may surprise you. The reason is that practically all the salt (NaCl) we eat ends up in our urine. Since this ends up in sewage and the concentration of Cl^- is not affected by sewage treatment, high levels of Cl^-

in a water sample is often an indication of the presence of sewage effluents. You might test for heavy metals or phenolic compounds if the water was being polluted with industrial effluents or sewage containing industrial wastes.

Suspended solids can be determined by filtering a known volume of water through a clean dry Whatman GF/C grade glass fibre filter. The filter is cleaned, dried at 105°C and weighed. 100 cm^3 of water is passed through the filter which is then dried at 105°C and reweighed. The dry mass of suspended solids can then be found by subtraction and is usually expressed as mg dm^{-3}. Initial testing might also look at some physical aspects of the water, for example colour, odour and turbidity. Water odour can indicate the presence of some substances, for example ammonia. Colour depends on both the suspended and dissolved solids. Thus organic material can impart yellow, greenish and brownish colours to water. The Hazen scale is a widely used method for colour grading. One Hazen unit equals the colour produced by dissolving 1 g m^{-3} of platinum in 2 g m^{-3} of cobaltous chloride. Turbidity gives a rough indication of the amount of undissolved material in the water and so is an important initial test. It is measured by comparing the samples with solutions containing specific concentrations of a dye called **formazin**, in **formazin turbidity** units.

Biological tests

The great cholera outbreak in London in the 1850s shows the importance of monitoring water for the presence of pathogenic bacteria, viruses and other microorganisms. The organisms most commonly used as indicators of biological quality are the **coliform** group of bacteria as a whole and *Escherischia coli* in particular. Coliforms are non-pathogenic but their presence in a water sample can readily be detected by biochemical tests (Fig 15.3). Pathogenic bacteria are more difficult to isolate and identify.

E. coli is a common, normal and harmless constituent of your gut flora. Since it passes out of your body with your faeces the presence of *E. coli* in a water sample indicates that:

- the sample is contaminated with human faecal material probably sewage effluent;

- the possible presence of pathogenic bacteria that also occur in your gut as well.

The absence of significant numbers of *E. coli* from a sample shows that the chance of faecal contamination of the water sample, and therefore the chance of faecal pathogens being present, is negligible.

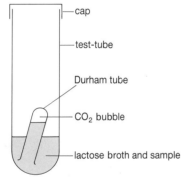

Fig 15.3 The presumptive coliform test for bacteria like *Escherischia coli*. The water sample is added to lactose broth. Only coliforms have the ability to ferment lactose to produce carbon dioxide which collects at the top of the Durham tube.

- cap
- test-tube
- Durham tube
- CO$_2$ bubble
- lactose broth and sample

QUESTIONS

15.1 A routine water sample contains high levels of chloride and coliform bacteria. What is the likely pollutant? Explain your answer.

15.2 (a) Investigate and write a report on the causes of the London cholera outbreak of the 1850s.

(b) Would finding large numbers of coliform bacteria in a water sample taken from a river necessarily mean that drinking water abstracted from the river would also be contaminated?

15.2 MEASURING OXYGEN DEMAND

Oxygen is essential to the survival of all aerobic organisms. So, clearly, the amount of oxygen present in a stream or lake is a key environmental variable for the animals and plants living there.

Measuring dissolved oxygen

Two techniques are available.

1. **The Winkler method**: This is a chemical method in which the oxygen present in the sample is replaced by its equivalent in free iodine. The amount of free iodine can then be measured by reacting it with sodium thiosulphate. The amount of thiosulphate used is then a measure of the amount of oxygen present in the water sample.
2. **Oxygen electrode** (see Fig 13.2): Here the oxygen concentration is measured by setting up a potential difference between two calibrated electrodes. This potential difference is then amplified and displayed on a meter. The size of the potential difference is directly proportional to the amount of dissolved oxygen.

Biochemical oxygen demand

The organic material present in an effluent will be oxidised by micro-organisms using up dissolved oxygen. Effluents can therefore be classified according to how much oxygen they require for the complete breakdown of the biodegradable organic material they contain. This is the basis of the 5-day biochemical oxygen demand (BOD) test (usually abbreviated to BOD_5). Technically, BOD_5 is defined as:

> The mass of dissolved oxygen, in grams per cubic metre or milligrams per cubic decimetre, taken out of solution by a water sample incubated in darkness at 20°C for five days.

I will refer to this test as BOD_5^{20} from now on. The method involves collecting two matched water samples A and B from the same site at the same time. (If the samples are suspected of containing high levels of biodegradable pollutants they may be mixed with well-oxygenated water at this stage.) The dissolved oxygen content of sample A is determined immediately. That of sample B is determined five days later after incubation for 20°C in the dark. During this time oxygen is consumed by microbial respiration as the bacteria break down the organic matter in the sample. BOD_5^{20} is then the difference in the dissolved oxygen content between sample A and B. Note that the temperature of incubation is critical. A BOD_5 is two times as great if the sample is incubated at 25°C as at 5°C.

The BOD_5^{20} test is essential for estimating the amount of biodegradable organic material in an effluent and is therefore a mandatory water quality test. However, it does not always give an accurate representation of the amount of biodegradable organic material present since it depends on the activity of microorganisms whose behaviour is not always consistent. For example, BOD_5^{20} determinations can be affected by the following.

- The presence of substances poisonous to microorganisms, e.g. heavy metals.

- Inorganic reducing agents like sulphide. Such substances give the water sample a chemical oxygen demand (COD).

- The availability of dissolved oxygen and nutrients.

- BOD_5^{20} does not estimate the oxygen demands of compounds which are difficult to degrade, e.g. cellulose. Thus the effluent discharged from a paper mill, which will contain a large amount of cellulose, may have a BOD_5^{20} which is less than that of sewage. However, its total BOD may be much greater.

- BOD_5^{20} does not measure the microbial activity which occurs after five days which may make a large contribution to total BOD. This consists predominantly of the growth of *Nitrosomonas* ($NH_3 \rightarrow NO_2$) and

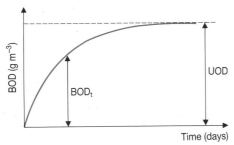

Fig 15.4 The relationship between the biochemical oxygen demand BOD and the ultimate oxygen demand (UOD).

Nitrobacter ($NO_2 \rightarrow NO_3$). These problems can be overcome using a more sophisticated test where oxygen consumption is measured over a short period under optimal conditions. It is then assumed that the total oxygen demand of the effluent will increase with time as shown in Fig 15.4 giving the ultimate oxygen demand (UOD). Often, BOD_5^{20} values are only half of the UOD.

ANALYSIS

'The dirty dozen'

This exercise will give you practice in answering exam questions.

Table 15.2 shows information about effluent discharges from industry in the Severn-Trent Water Authority Area in 1974. The eight factories were amongst twelve designated as 'the dirty dozen' because their effluents were so far below the quality standard set by the Water Authority.

(a) (i) Suggest what is meant by 'consent effluent quality'.

(ii) Suggest what is meant by 'mean effluent quality'. State how, in principle, such figures may be obtained.

(iii) Explain how the figures for 'suspended solids' are obtained.

(b) From the table, work out which factory on average is adding most to

(i) the BOD of the river into which it is discharging, and

(ii) the suspended solids.

(c) (i) Explain how the discharge of hot water (e.g. factory F) may affect the organisms present in the river.

(ii) Explain briefly what other information you would require in order to assess the effects of a proposed rate of effluent discharge into an unpolluted river.

Table 15.2

Factory	Volume of effluent (megalitres per day)	Consent effluent quality (mg dm⁻³)		Mean effluent quality (mg dm⁻³)	
A·	14	BOD	20	BOD	55
B	3	SS	30	SS	47
C	1	BOD	20	BOD	53
		SS	30	SS	107
D	0.03	SS	30	SS	672
E	3	SS	100	SS	124
F	2	BOD	20	BOD	28
		SS	30	SS	102
		Temp.	25°C	Temp.	28.2°C
G	0.2	SS	50	SS	789
H	0.8	BOD	29	BOD	33
		SS	30	SS	34

SS = suspended solids; BOD = biochemical oxygen demand.

Chemical tests of oxygen demand

Industry cannot wait five days for a BOD_5^{20} test before discharging its effluent. As a result there are a range of chemical tests available, whose exact relationship to the BOD_5^{20} for a particular effluent is known, which are used to estimate oxygen demand of the effluent. We will examine two of these. Do not worry too much about the chemical details; it is the principles which count.

Permanganate value

This test measures the amount of material oxidised by the strong oxidising agent potassium permanganate. The sample is incubated at 27°C for four hours. The relationship between the four-hour permanganate value (PV) and the BOD_5^{20} for sewage is 1:3.5. Thus a sewage effluent which uses 100 $g m^{-3}$ of oxygen on the PV scale will require 350 $g\ m^{-3}$ on the BOD_5^{20} test.

However, there are many chemicals which are oxidised by permanganate but which have no BOD because they are not decomposed by micro-organisms, e.g. iron II salts, nitrites, thiosulphates, oxalic acids and phenols. These substances, which have a chemical oxygen demand, interfere with the measurement of the oxygen demand of the bulk of the biodegradable organic material in the sample. To overcome this problem another PV test which only lasts three minutes is performed. During this time the substances listed above which require chemical oxidation, e.g. iron II salts and so on, will be oxidised by the permanganate. This three-minute PV test is used with effluents which do not contain much organic material, e.g. from mine drainages and from coking plants (Fig 15.5).

Interestingly, the ratio between the four-hour PV value and the three-minute PV value gives some indication of the origin of the polluting material (Table 15.3).

Fig 15.5 White hot coke emerging from a coking oven. This process, which involves heating coal to a high temperature, produces particularly obnoxious gases. These are dissolved in water to produce an effluent.

Table 15.3   Permanganate (3-minute and 4-hour) values and the 4-hr/3-min PV ratio.

Liquid	Permanganate value		Ratio
	3-min	4-hr	4-hr/3-min
sewage effluent	2.6	7.8	3.0
pea washing wastes	688	3000	4.4
potato washing wastes	400	2300	5.8
carrot canning wastes	151	1450	9.6
brewery wastes	320	2080	6.5

Chemical Oxygen Demand

This test, which uses potassium dichromate rather than potassium permanganate as an oxidising agent, effects a more complete oxidation of organic matter than does the PV test. Therefore the chemical oxygen demand (COD) value is often preferred for assessing organic pollution. Since the COD test is rapid (taking about two hours) it is often used for the routine monitoring of an effluent once its BOD–COD ratio is known. Examples of the strength of various effluents as measured by BOD_5^{20}, PV, and COD tests are shown in Table 15.4.

Table 15.4   The strength of various industrial effluents.

Type of waste	Main pollutants	BOD (5 days/g m^{-3})	PV (4 hours/g m^{-3})	COD (g m^{-3})
abattoir	suspended solids – protein	2 600	550	4 150
beet sugar	suspended solids – carbohydrate	850	80	1 150
cannary (meat)	suspended solids – fat, protein	8 000	1 520	17 940
chemical plant	suspended solids – organic chemicals	500	370	980
coke ovens	phenols, cyanide	780	1 420	1 670
gas works	thiocyanate, thiosulphate	6 500	12 200	16 400
smokeless fuel	ammonia	20 000	27 500	

15.3 (a) Why is it essential in a BOD test to incubate the second sample in the dark?

(b) When could an industrial effluent show a low BOD_5^{20} and yet contain a large amount of organic matter? Give an example of such an effluent.

(c) Calculate (i) BOD–PV ratios (ii) BOD–COD ratios for the effluents listed in Table 15.4

(d) Can you distinguish two general types of waste on the basis of their BOD–PV ratios? Account for any distinction that you have made.

(e) If the oxygen demand of an effluent exceeds the oxygen supply from the receiving water what will happen?

15.3 BIOLOGICAL MONITORING

The basic idea behind biological monitoring is that clean, unpolluted water contains a characteristic assemblage of organisms, including stonefly nymphs (larvae), *Gammarus pulex* (the freshwater shrimp), mayfly nymphs and the larvae of caddis flies. These are **indicator species** for clean water and a community of animals which contains these species will be an **indicator community** for clean water. Organic pollution, for example, will change this community. Species which are sensitive to organic pollution will disappear or their numbers will be reduced. Species tolerant of organic pollution, for example blood worms, rat-tailed maggots and tubificid worms, will become more common. If we can measure and quantify these changes then we can use organisms to monitor pollution. The organisms most commonly used are macro invertebrates.

Advantages of biological monitoring include:

- Detecting intermittent pollution. For example, consider a chemical plant discharging, say, copper into a stream. The discharge occurs once a week first thing on a Tuesday morning. You take your water sample for chemical analysis on a Friday morning. By then the stream has washed away the contaminant and your sample does not show the presence of copper. However, the animals present in the stream will be affected by the discharge. Some species will disappear so the composition of the community changes. This change will be detectable for weeks, perhaps months.

- Chemists cannot hope to analyse routinely a water sample for the 1500+ known pollutants. Biologists, however, may detect changes in the aquatic community which suggest that pollution is occurring. The chemists can then begin a screening procedure to look for the pollutants producing the observed change. For example, in 1969 a large number of sea birds died unexpectedly in the Irish Sea. The birds contained large amounts of PCBs. This biological early warning provided the first clues that (i) chemical industries were discharging effluents containing PCBs into inshore water (ii) PCBs could damage this environment.

- Chemicals which occur in low concentrations in the water may be accumulated in the bodies of organisms. For example, in one study, water in Lake Windermere had a lead concentration of 45 μg dm^{-3} ($\equiv$ 45 ppb) while the phytoplankton had a lead concentration of 278 mg g^{-1} dry weight, a concentration factor of some 5000. Such accumulation could be useful in detecting the presence of pollutants in water where the concentration in the water itself is too low to be measured with available technology.

Disadvantages include:

- The natural variation in faunal assemblages. For example, stonefly nymphs (larvae) play an important part in biotic indices designed to monitor organic pollution (see later). However, these animals are not abundant in lowland rivers. This is not because such rivers are polluted but rather that stoneflies prefer upland rivers rich in dissolved oxygen.

- Species which may be sensitive to one pollutant may be tolerant of another. For example, stonefly nymphs cannot tolerate organic pollution but can tolerate quite high concentrations of heavy metals. So the presence of a species does not confirm that the water is totally unpolluted. Rather it tells us that the water is not polluted in a particular way.

- Macro invertebrates show an incredibly patchy distribution in stream beds. This means that one sample may collect very few invertebrates while another may collect a lot. If only one or two samples are taken at each site then some groups may be under-represented simply by chance, not because they are not present.

- The need for taxonomic expertise on the part of the operators.

Trent biotic index

This index of water quality was developed in 1964 for use by the Trent River Authority. While it has now been largely superseded by other, more powerful, biotic indices its ease of use makes it convenient for us to study. The basic concept underlying this, and other biotic indices, is that stream animals disappear following pollution by biodegradable organic pollutants in the order shown in Fig 15.6. Thus stoneflies are most sensitive to organic pollution like sewage while oligochaetes are the least sensitive. In fact, the key variable is the reduction in oxygen levels in the water following pollution by biodegradable organic pollutants.

1 Stonefly nymph
(up to 30 mm)

2 Flattened mayfly nymph
(up to 16 mm)

3 Caseless caddis fly larva
(up to 26 mm)

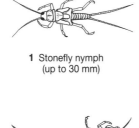

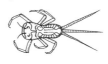

4 Swimming mayfly nymph
e.g. *Baetis rhodanii*
(up to 11 mm)

5 Freshwater nymph
(up to 20 mm)

6 Water louse
(up to 12 mm)

7 Blood worm (or midge larva)
(up to 20 mm)

8 Rat-tailed maggot
(up to 55 mm inc. tube)

9 *Tubifex tubifex*
(up to 40 mm)

Fig 15.6 The effect of organic pollution on different invertebrates. Species are affected in the order shown by the numbers. Stoneflies are very sensitive; annelids like *Tubifex* and rat-tailed maggots are the least sensitive.

To calculate the Trent biotic index:

- Collect macro invertebrates with a hand net using 'kick' (Fig 15.7) and hand sampling from the stream.

- The animals collected are then identified to the appropriate level, either species, genus, family or even higher taxa (Table 15.5). Thus in your

Fig 15.7 A freshwater biologist taking a kick sample from a stream. Animals dislodged from the stream bed by vigorous kicking are swept by the current into the net.

Table 15.5 The Trent biotic index.

The term 'Group' used for purpose of the biotic index means any one of the species included in the following list of organisms or sets of organisms.

Each known species of Plathyhelminthes (flatworms)
Annelida (worms excluding genus *Nais*).
Genus *Nais* (worms)
Each known species of Hirudinea (leeches)
Each known species of Mollusca (snails)
Each known species of Crustacea (hog louse, shrimps)
Each known species of Plecoptera (stone-fly)
Each known genus of Ephemeroptera
(may-fly, excluding *Baetis rhodanii*)

Baetis rhodanii (may fly)
Each family of Trichoptera (caddis-fly)
Each species of Neuroptera larvae (alder fly)
Family Chironomidae (midge larvae except *Chironomus Ch. thummi*)
Chironomus Ch. thummi (blood worms)
Family Simulidae (black-fly larvae)
Each known species of other fly larvae
Each known species of Coleoptera (beetles and beetle larvae)
Each known species of Hydracarina (water mites)

Biogeographical region: Midlands, England		Total number of groups present									
		Biotic Indices									
		0–1	2–5	6–10	11–15	16–20	21–25	26–30	31–35	36–40	41–45
Plecoptera nymphs present	More than one species	–	7	8	9	10	11	12	13	14	15
	One species only	–	6	7	8	9	10	11	12	13	14
Ephemeroptera nymphs	More than one species*	–	6	7	8	9	10	11	12	13	14
	One species only*	–	5	6	7	8	9	10	11	12	13
Trichoptera larvae present	More than one species†	–	5	6	7	8	9	10	11	12	13
	One species only†	4	4	5	6	7	8	9	10	11	12
Gammarus present	All above species absent	3	4	5	6	7	8	9	10	11	12
Asellus present	All above species absent	2	3	4	5	6	7	8	9	10	11
Tubificid worms and/ or Red Chironomid larvae present	All above species absent	1	2	3	4	5	6	7	8	9	10
All above types absent	Some organisms such as *Eristalis tenax* not requiringdissolved oxygen may be present	0	1	2	–	–	–	–	–	–	–

* *Baetis rhodanii* excluded
† *Baetis rhodanii* (Ephem.) is counted in this section for the purpose of classification

sample you need to identify each known species of flatworm (Platyhelminth) but simply identify worms as being annelids, with the exception of members of the genus *Nais* which need to be identified separately.

- After identification you then count the total number of groups you have identified from your sample, say 15. You then turn to Table 15.5 and work down the column headed 11–15 (in bold in the table) answering the questions on the left. Thus if your sample contained two or more species of Plecoptera (stonefly nymphs) then the biotic index is 9. If no stoneflies and no Ephemeroptera (mayfly) nymphs were present but you found one species of Trichoptera (caddis fly) nymph then the Trent biotic index is 7.

- Clean streams score an index close to 10, dropping to lower values with increasing organic pollution and therefore lower oxygen content, reaching an index of 1–2 for heavily polluted streams.

Note that the Trent biotic index does not take into account the relative abundance of the organisms involved. So simply one stonefly nymph in a sample will be given the same weight as twenty. Currently, the National Rivers Authority uses a more complicated but better biotic index called the Biological Monitoring Working Party (BMWP) score. This uses more taxonomic groups than the Trent biotic index and so requires a greater level of taxonomic skill to use. Nonetheless, it gives more robust results than the Trent biotic index.

ANALYSIS

Comparing sites using the Trent biotic index
This exercise requires you to classify sites as being polluted or unpolluted using the Trent biotic index.

Table 15.6   Invertebrates found in kick samples taken from three stations in a stream. Station 1 is the highest sampling point on the river, station 3 the lowest.

Species		Station		
		1	2	3
Baetis rhodani	(Eph)	15	12	3
Ecdyonurus sp.	(Eph)	3	1	
Rhyacophila sp.	(T)		4	
Hydropsychidae sp.	(T)	1	1	
Gammarus pulex	(C)	8	8	
Asellus aquaticus	(C)	14	14	17
Chironomus sp.	(Ch)	37	37	37
Limnaea peregra	(M)	1	1	1
Dicranota sp.	(F)	1	1	
Tubifex tubifex	(Tub)	28	26	9
Erpobdella sp.	(L)	1	1	1
Hydrachna sp.	(Hyd)	7	1	
Nais sp.	(A)			42

Eph = Ephemeroptera, T = Trichopter, C = Crustacean, M = mollusc, F = other flies, Tub = Tubificid, L = leech, Ch = Chironomid, Hyd = Hydracarina, A = annelid.

Table 15.6 gives some data on invertebrates found in kick samples taken from three stations in a stream.
(a) Calculate the Trent biotic index for each of the stations.
(b) Which station is the heaviest polluted and which the least?
(c) What sort of material could be polluting the streams?
(d) Where does it enter the stream?

Diversity indices

This alternative approach to biotic indices is based on the assumption that as the pollution load in a stream increases there is a reduction in the total number of species present but an increase in the number of individuals in species which are tolerant of pollution. This leads to the idea that measuring the species diversity of the community in a stream in some way measures the 'pollution stress' being suffered by the stream. The higher the stress the lower the species diversity. Again, there are a number of techniques for measuring diversity, all of which rely on mathematical relationships between the number of species (S) and the abundance of individuals (N).

Shannon-Weiner diversity index
This index, H, which is derived from information theory, tries to describe the diversity of a community in terms of the amount of information it

contains. This is quite a difficult concept to understand but think of it like this. A complex community will require a bigger book to describe it, i.e. it contains more information, than a simple community. The formula looks a bit daunting but with a little practice you will soon get used to it.

$$H = -\sum_{i=1}^{s} (p_i)(\log_e p_i)$$

where S is the number of species and p_i is the proportion of individuals in the ith species. A worked example (Table 15.7) will show how this index can be calculated.

Table 15.7 Calculating the Shannon-Weiner diversity index.

	Species A	Species B	Species C	Total
No. of individuals	25	32	43	100
p_i	25/100 = 0.25	32/100 = 0.32	43/100 – 0.43	1
$\log_e p_i$	–1.39	–1.14	–0.84	
$(p_i)(\log_e p_i)$	0.25 × –1.39 = –0.35	0.32 × –1.14 = –0.36	0.43 × –0.84 = 0.36	

$H = - [(-0.35) + (-0.36) + (-0.36)] = 1.07$

ANALYSIS

Using diversity indices

This exercise will give you practice in using the techniques discussed so far in this chapter.

Table 15.8 shows the numbers of organisms found at two stations on each of two rivers (A and B). Conditions were similar at all four stations with the exception that A_2 and B_2 were samples below the point of entry of polluting discharges.

(a) Describe briefly how an investigator would have obtained the data in Tables 15.8 and 15.9.

(b) Explain what the figures in both tables indicate about the nature of the pollution at A_2 and B_2.

(c) (i) Give one example of a simple 'diversity index' and calculate it for each of the four stations in Table 15.8.
(ii) Explain how your calculated values correlate with your answer to (b).

(d) Suggest what other measurements could be made at station B_2 to confirm the nature of the pollution.

Table 15.8

	Mean number of organisms per m² of river bed			
Organisms	A_1	A_2	B_1	B_2
Ancylus	10	–	8	–
Asellus	15	489	17	9
Baetis	57	–	52	3
Chironomus	2	241	2	5
Ecdyonurus	21	–	19	2
Elmis	9	–	11	2
Gammarus	38	9	31	1
Hydracarina	35	–	39	–
Hydropsyche	4	–	7	4
Limnaea	8	–	11	–
Perla	16	–	12	1
Polycelis	17	–	13	–
Tubifex	–	189	–	–

Table 15.9

	A_1	A_2	B_1	B_2
dissolved oxygen mg dm^{-3}	11.2	4.1	11.4	12.1
BOD mg dm^{-3}	1.0	5.8	1.8	2.0
suspended solids mg dm^{-3}	2.2	6.9	2.3	1.8

Biochemical monitoring

When exposed to certain types of pollutant some organisms start to produce large quantities of some proteins. For example, fish exposed to substances like polycyclic aromatic hydrocarbons show increased levels of an enzyme complex called cytochrome P_{450} oxidase in their liver. This happens because the pollutant induces the genes which encode the enzyme, causing more enzyme to be produced. Similarly, many freshwater organisms produce a protein called metallothionein in response to exposure to heavy metals.

In theory, then, catching a sample of fish, removing their livers and testing the level of cytochrome P_{450} oxidase activity or metallothionein in them should provide an easy means of monitoring pollution by aromatic hydrocarbons, PCBs and heavy metals. Unfortunately, the biochemistry is currently a little difficult but we can perhaps envisage a time when a simple test kit might be available.

QUESTIONS

15.4 (a) What would the biotic index of pure water be?
   (b) Why will some organisms, like annelids and chironomids, be more tolerant than others of organic pollution?
   (c) Would it be fair to conclude on the basis of a score of 10 on the Trent biotic index that a river was unpolluted?
   (d) In what ways are diversity indices similar to and different from biotic indices?

15.5 Why do you think biochemical tests could prove to be particularly useful in monitoring pollution?

SUMMARY ASSIGNMENT

1. How would you determine the dissolved oxygen content, BOD and dissolved solids in a water sample?

2. (a) With reference to named invertebrates, explain what is meant by 'indicator species' and 'indicator communities' and discuss their use in monitoring water quality.
   (b) Compare and contrast the chemical and biotic (biological) methods of monitoring water quality.

3. Table 15.10 shows the numbers of different organisms found at two locations (A and B) on the same river. The two locations were similar except for the presence of organic pollution at one of them.
   (a) Explain the following: (i) biotic index (ii) indicator species (iii) diversity index.
      Apply each term to the information in the table, making calculations where appropriate.
      What conclusion can you draw about the nature and extent of the pollution?
   (b) Account for the occurrence of a named organism (i) present only at A (ii) present only at B (iii) present at both A and B.

Table 15.10

Organism	Numbers per m² of river bed	
	A	B
Baetis	60	–
Ecdyonurus	19	–
Asellus	15	512
Chironomus	–	229
Gammarus	40	4
Hydropsyche	4	–
Tubifex	–	207
Polycelis	15	–
Limnaea	8	–
Ancylus	6	–
Hydracarina	37	–
Elmis	9	–
Perla	14	–

4. Start collecting newspaper cuttings on pollution incidents. These will provide invaluable, up-to-date information which you can use to illustrate your exam answers.

Appendix A

ANSWERS TO ANALYSIS EXERCISES

CHAPTER 2
The effects of turbulence

(a) (*i*) 12.5 m. (*ii*) 45 m. The thermocline represents the bottom of the mixed layer of surface water. Its presence and the depth at which it occurs depends on the amount of turbulence in the water column and the intensity of heating of the surface waters. The compensation depth depends on the clarity of the water and the intensity of the sunlight. Here there is sufficient light penetration for the compensation depth to lie below the thermocline.

(b) Nitrate is used by photoautotrophs in the surface waters and regenerated by decomposition in deeper water. However, the thermocline prevents this regenerated nitrate from returning to the surface water. Hence, nitrate levels are depleted in the surface water but higher in deeper waters where most of the decomposition is occurring.

(c) (*i*) No (*ii*) No (*iii*) Possibly.

(d) (*i*) Wind induced turbulence, i.e. wave action, has increased, so the depth of the mixed layer has also increased. (*ii*) Nitrate regenerated in deep waters has been brought to the surface by the increased turbulence. (*iii*) It will break down so that the water temperature becomes more similar throughout the water column. (*iv*) It should increase it providing the mixing is not sufficiently strong to carry photoautotrophs below the critical depth.

(e) Mixing is now very strong so photoautotrophs are probably being carried below the critical depth. As a result NPP will fall because cells are now becoming light limited.

CHAPTER 4
The rate of growth of a fish stock

(a) You are given years 1 to 4. Year 5 = 52, year 6 = 78, year 7 = 142, year 8 = 262.

(b) Your graph should show an exponential increase in stock biomass with time.

(c) Plot stock biomass at S_n on *x*-axis against stock biomass at S_{n+1} on *y*-axis, i.e. for year 5, $S_n = 52$, $S_{n+1} = 78$. Again you get an exponential curve showing the stock is growing at a faster and faster rate.

(d) We are assuming that there is no competition between fish for food, mates and so on. In reality the stock growth curve would level out to produce an S-shaped growth curve.

Surplus yield depends on stock size

(a) To do this you need to work out the natural mortality at each stock density and then calculate how many fish you can catch during the year so that the stock biomass is the same as it was at the beginning of the year. For example, take a stock biomass of two units. Natural mortality is zero and the stock increases to six units of biomass. So you can take four. Now with a stock size of 14 we begin to have to take into account natural mortality, which equals six units of biomass. The stock only increases to 22 units so you can catch only two. The figures in sequence are 4, 8, 2, 24, 14, 40, 78.

(b) Plot initial stock biomass on the *x*-axis and surplus yield on the *y*-axis.

(c) Generally the trend is upwards but you get troughs at certain initial biomasses. Why do you think this happens?

(d) Providing abiotic conditions do not change, which they probably will, yes.

Using egg counts to estimate stock size

(a) $3.5 \times 10^{12} / 70\,000 = 5 \times 10^7$

(b) 10^8

(c) About half as large. Are you surprised by this difference?

Interpreting surplus yield curves

(a) Fishing effort which is directly proportional to stock size.

(b) They have simply plotted it by eye through the points. Data on catch and fishing effort would have been taken from records kept at ports.

(c) 95×10^6 lb yr^{-1} and 180×10^6 lb yr^{-1} respectively.

(d) 9.5×10^3 lb yr^{-1} annual standard fishing day^{-1}; 7.2×10^3 lb yr^{-1} annual standard fishing day^{-1}.

(e) 35×10^3 annual standard fishing days. 5.1×10^3 lb yr^{-1} annual standard fishing day^{-1}.

CHAPTER 6
Putting all the measures together

(a) See text.

(b) See text.

(c) (*i*) Your answer should refer to crop development. Firstly the increase in LAI as the individual plants grow and produce more leaves per unit area of ground. Secondly the decrease as the crop matures and the leaves begin to senesce and die. (*ii*) Your answer needs to refer to the time of planting which relates to the frost hardiness of crops or the need to avoid vernalisation (defined in the next chapter) in the case of sugar beet.

(d) Remember CGR = LAI × NAR. So the more leaves there are per unit land area the greater the crop productivity up to a certain point when LAI can exceed an optimum value. Now lower leaves are being shaded and may be below their compensation point and NAR therefore decreases.

(e) (i) Both crops have maximum NAR in June (why?) but do not reach maximum LAI until August/September when NAR is falling (again you need to explain why). So CGR is not at a maximum because of the mismatch between maximum NAR and LAI. Planting earlier would mean the plants would achieve their maximum LAI earlier, i.e. at higher rates of NAR, so increasing CGR and hence productivity. (ii) Neither crop is frost hardy and sugar beet vernalises.

(f) CGR = LAI × NAR = 1.5 × 40 = 60 g m^{-2} week^{-1}

(g) You should consider factors like light intensity, temperature, leaf area and age, growth stage of the plant e.g. is it putting more resources into producing tubers or grain, senescence.

CHAPTER 7
Worlds apart

You need to find out about sea currents and wind directions for the two areas. Then using the information presented so far in this chapter and your brain, i.e. thinking about the problem, you should be able to produce a coherently argued explanation. Ask you tutor and friends for help if you get stuck.

The importance of hedges

(a) (i) Wind speed and evaporation from the soil. (ii) 10 and 7 m respectively. (iii) The air temperature is increased nearer to the hedge. Warm air can hold more moisture than cold air, hence relative humidity increases.

(b) (i) Possible reasons could include competition from the roots of the hedge plants, shading by the hedge, increased fungal disease due to higher relative humidity, insect pests immigrating on to the crop from the hedge. (ii) The reduction in wind speed, combined with significantly higher air temperatures and soil moisture, means that the plants are able to photosynthesise more efficiently, are less likely to lodge, in the case of cereal crops, and will be less likely to suffer water deficiency.

(c) (i) A reduction in yield. The increase in crop yield (the area under the crop yield curve) from 2H to 18H from the hedge is greater than the losses close to the hedge. (ii) To improve access and efficient use of large machines. To remove reservoirs for pests and diseases (more about this in chapter 10).

CHAPTER 8
The answer lies in the soil

Consult your tutor to discuss your answers.

The water-holding capacity of different soils

(a) 2%, 8.5%, 10.9%, 13.5% respectively if you exclude gravitational water.

(b) They contain more pore space because of the small size of the clay particles.

(c) Water is held more tightly in the small pores of clay soils, i.e. more water is unavailable to the plants, compared to water held in the larger pore spaces of sandy soils.

CHAPTER 9
Fertilisers and edaphic factors

(a) Decomposition of grass ploughed into the soil to humus improves the soil structure and, therefore, its aeration and water-holding capacity, whilst also supplying nutrients. Both factors will promote the growth of barley. See text for further details.

(b) Plants in 1976 were water limited compared to plants in 1977.

(c) This confirms (b) since boulder clay soils will retain more water than limestone soils.

(d) The grass needs to be kept growing throughout the year so its needs regular feeding. If all the nitrogen were applied early in the season then much would be lost through leaching before it could be used.

Economics and fertiliser application

(a) Plot N applied on x-axis and yield on the y-axis. You should get an asymptotic curve which levels out at higher levels of N application.

(b) The increased cost of the fertiliser is £25 for each increase in application rate. The increase in yield for each application is, respectively, £50, £90, £50, £10 giving benefit cost ratios of 2, 3.6, 2 and 0.4.

(c) No – at the highest rate of application costs exceed benefits.

(d) They reach a peak when N is applied at a rate of 100 kg ha^{-1} and then fall off.

(e) No.

(f) 100 kg ha^{-1} N applied.

Crop rotation

(a) Barley is a gross feeder so it can make the best advantage of nutrients released from decomposing animal dung dropped on the field in the winter.

(b) Red clover is a legume so it will fix nitrogen and boost soil nitrogen levels.

(c) Any addition of organic material improves soil structure (see text for details) whilst decomposition of the grass would release valuable nutrients.

(d) Animals provide a source of nutrients in their dung which is easy for plants to access whilst also improving soil structure.

(e) Nutrients would be locked into the turnips rather than be leached away by the winter rains.

(f) Intensive systems of farming must keep fields producing the most valuable crops, so you have to lose the animals stages of the rotation for economic reasons.

(g) Economics. The continual application of inorganic fertilisers combined with a lack of organic input will lead to an impoverished soil structure and soil acidification. See text for further details.

CHAPTER 10
Intraspecific competition

(a) Light, water, nutrient availability, temperature and pest infestation must be the same for all pots. Alternatively replicates of each pot must be arranged at random on, say, the bench of a green house, to prevent systematic bias in any of the factors.

(b) Plants are competing for available nutrients, water and light which leads to a reduction in yield per plant.

(c) 238 g m^{-2}, 270 g m^{-2}, 329 g m^{-2}. Whilst sowing at low density may increase yield per unit plant, sowing at higher densities may increase overall yield because more of the available resources are being used.

CHAPTER 11
The benefits and costs of using fungicides

(a) No. This could be due to changing levels of fungal infection as a result of different abiotic factors either affecting the fungus directly or by affecting the plant.

(b) No. Different varieties will vary in their susceptibility to fungal attack and so some will respond better than others to application of fungicide.

(c) This graph shows the build up of resistance in a fungal population. Too much spraying with the same fungicide causes this effect.

Pulling the threads together

All the information needed to answer the questions is in the text but spread over several chapters. Use this as an opportunity to review your progress and knowledge. Ask your tutor or friends if you get stuck.

CHAPTER 12
Defining pollution

(a) The stream may be full of rubbish but it actually supports a varied and thriving community of plants and animals.

(b) A sewage outfall pipe. If the sewage has been properly treated then this effluent is harmless.

(c) This lake is badly acidified and is probably the most damaged of all the habitats you have seen. Note the badly acid-damaged tree in the foreground.

A complex problem

(a) Dieldrin, used as a seed dressing, has probably leached from the fields in to the lake.

(b) During the winter the thermocline breaks down so that the hypolimnion becomes oxygenated. This changes the redox conditions of the bottom sediments which release cadmium into the water.

(c) Mercury has been converted to dimethyl mercury by anaerobic bacteria at the bottom of the lake. This compound has then entered the fish either directly from the water or indirectly via their food.

(d) Could be a synergistic effect of the pollutants acting together and compounded by low levels of oxygen in the lake due to the high summer temperatures.

CHAPTER 13
Dissolved oxygen and temperature

(b) As the temperature rises the concentration of dissolved oxygen falls.

(c) We often associate an increase in temperature with an increase in the solubility of a substance.

(d) Such discharges will reduce oxygen levels in the receiving water. If the reduction is large enough then aerobic organisms might die. If the stream is already polluted with biodegradable materials which exert a BOD the loss of oxygen could be catastrophic.

(e) A dramatic fall in oxygen concentration as a result of increased temperature and the oxidation of sulphide.

Pollution in the River Trent

(a) The increase and subsequent fall in BOD and the characteristic changes in the organisms below the outfall pipe. See text for more details which you can use to explain these changes.

(b) Tubificids.

(c) Changes in the level of oxygen in the water. See text for further details.

(d) Low oxygen concentration and high BOD.

(e) Both animals contain haemoglobin which allows them to gather oxygen from the water and store it even when oxygen is at a low concentration; remember haemoglobin has a high affinity for oxygen.

CHAPTER 14
The origin of acid rain

The decrease in pH of rain water is accompanied by an increase in both nitrate ions (from nitric acid) and sulphate ions (from sulphuric acid). There is a reduction in bicarbonate ions. These increases in the concentration of sulphate and nitrate ions suggest that the source of the increased acidity is the burning of fossil fuels. Explain their origins from information given in the text.

Eutrophication

(a) See text.

(b) Using a Secchi disc, a black and white disc of known diameter which is lowered in to the water. When you can no longer see the disc you measure the distance you have lowered the disc into the water which gives you a comparative method for estimating light penetration in different water bodies.

(c) Chlorophyll concentration is a measure of the number of autotrophic cells in the water which affects turbidity and so light penetration. The higher the chlorophyll concentration the smaller the light penetration.

(d) Faster in a eutrophic lake because of the increased amount of organic material. Hypolimnetic water may therefore become anoxic, leading to anaerobic processes and death of fish. See text for more details. $\mu g\ m^{-3}\ week^{-1}$.

(e) In summer all phosphorus is locked up in organic matter or attached to mud. Decomposition and reaeration of the hypolimnion in winter will release the phosphorus to the water column.

(f) Nitrogen, e.g. nitrate, levels increased.

(g) See text.

CHAPTER 15

The dirty dozen

(a) (*i*) The quality of effluent that the factory is allowed to discharge into a receiving water by the National Rivers Authority. (*ii*) The quality of the effluent over a number of sampling occasions. Regular sampling and analysis of the effluent. (*iii*) See text.

(b) (*i*) Factory A. (*ii*) Factory E.

(c) (*i*) See previous chapter. (*ii*) Volume of water and flow rate in the river. Previous discharges. BOD of receiving water. Toxicity of discharge.

Comparing sites using the Trent Biotic Index

(a) 8, 8, 6 respectively.

(b) 3 heaviest, cannot distinguish 1 and 2 basis of TBI.

(c) Biodegradable material like sewage.

(d) Between stations 2 and 3.

Using diversity indices

(a) See text. You need to explain both abiotic measurements and how organisms were collected.

(b) Briefly, eradication of oxygen dependent species at A_2 (give examples) and a dramatic increase in organism like *Chironomus* and tubificids which can survive at low oxygen concentrations. Looks like sewage. At B_2 general reduction in numbers of all species but oxygen lovers still present. Suggests the presence of a chronic toxic pollutant like a heavy metal or acid pollution (note the eradication of both species of mollusc, *Ancylus* and *Limnaea*) which does not affect oxygen concentration.

(c) (*i*) See text and follow calculation. (*ii*) You will see a reduction in species diversity at both sites, but a greater one at site A_2 compared to site B_2. This suggests that more groups, i.e. the oxygen-requiring species, are being eradicated by the pollution at A_2 compared to that at B_2.

(d) Tests for heavy metals, organic pollutants, acids etc. See text for details.

Appendix B

ANSWERS TO QUESTIONS

CHAPTER 1

1.1 (a) 2100 kPa
  (b) 40 100 kPa
  (c) 40 100 kPa
  (d) 108 730 kPa (Don't forget to add on the air pressure!)

1.2 (a) For organisms in the epipelagic zone, especially autotrophs, a major problem is to remain in the sunlit surface waters. So you might expect to see adaptations which aid floating or at least reduce sinking. For example, large surface area to volume ratios. Animals feeding on small autotrophic cells will have mechanisms for filtering the small cells out of the water.
  (b) The major problems in this deep water will be (*i*) the dark (*ii*) the lack of food. Animals will have adaptations for finding mates and food which include complicated light-generating devices. In some species of angler fish the male becomes parasitic on the female. Any passing meal must be taken so you find fish with enormous mouths which can consume food items several times larger than themselves. Invertebrates will feed either on detritus or on bacteria and fungi decomposing the detritus so filtering mechanisms are again common.

1.3 Cyanobacteria are prokaryotes but phytoplankton are all eukaryotes. Autotrophs are organisms which convert carbon dioxide into organic compounds; cyanobacteria fit into this category.

1.4 To answer the question properly, i.e. compare and contrast, you need to find similarities and differences in the role of zooplankton and microorganisms in marine ecosystems. Your answer should include reference to the role of zooplankton as herbivores, carnivores and detritivores. Give specific examples of each type. Microorganisms include bacteria, fungi and protocists like flagellates. Such organisms are both autotrophic and heterotrophic so you should give examples of both roles.

1.5 (a) (*i*) Diatoms and flagellates. (*ii*) Any of the animals which feed on the diatoms or flagellates. (*iii*) *Sagitta*, Sand eels, Hyperiid amphipods, herring.

  (b) 2.

1.6 (a) CO_2, H_2O
  (b) CO_2, H_2O
  (c) CO_2, H_2O, a source of nitrogen, e.g. nitrate, and possibly sulphur, e.g. sulphate.
  (d) CO_2, H_2O, a source of nitrogen, e.g. nitrate, and of phosphorus, e.g. phosphate.

1.7 (a) (*i*) Antarctic waters have high nutrient levels because of upwelling and so have high productivity, therefore lots of krill and lots of whales. (*ii*) The food previously consumed by the whales is now available for penguins and seals which increase in numbers.
  (b) Could slow down recovery and reduce penguin numbers if overfished.

1.8 Reduce primary productivity since autotrophs and nutrients would be removed from the surface waters by the downwelling, convergent currents.

1.9 (a) No, though a seasonal thermocline will develop in most summers.
  (b) By mixing warm surface water with cooler subsurface water.
  (c) No. The bottom of the euphotic zone depends on light availability not temperature.

CHAPTER 2

2.1 (a) NPP measures the amount of carbon fixed per unit time whilst the standing crop is only a measure of the amount of carbon present at a particular time. The rapid reproduction and growth rates of marine autotrophs account for the high rates of production.
  (b) Algal beds, reefs and estuaries. You should think about the availability of sunlight and nutrients.
  (c) Marine autotrophs have very high rates of productivity, therefore a low standing crop of autotrophs can support a larger standing crop of grazing zooplankters.

2.2 (a) (*i*) No (*ii*) No (*iii*) Yes.
  (b) Blue.
  (c) Enables them to trap a wider range of wavelengths of light.

2.3 (a) Initially photosynthetic production increases linearly as light intensity increases. However, at higher light intensities photosynthetic production levels off and then declines as light intensity increases further.
  (b) (*i*) Light availability. (*ii*) Availability of CO_2. (*iii*) This results from photo-inhibition probably due to the destruction, (bleaching) of chlorophyll at high light intensities.

2.4 (a) 22–23 m.
  (b) Strong sunlight is causing photo-inhibition.
  (c) Reduction in light availability.

2.5 (a) The compensation depth is the depth in the water column at which the compensation light intensity occurs.
  (b) Reduce it.

(c) (i) Basically same all year. (ii) Increase in summer, decrease in winter. (iii) Same as (ii).

(d) (i) The cell above the compensation depth will be photosynthesising faster than it is respiring so its NPP will be positive. The one below the compensation depth will have negative NPP.
(ii) Probably – to reduce their rate of sinking and so keep them in sunlit surface water.

(e) Yes.

2.6 (a) The greater the amount of phosphate the faster and greater the productivity.

(b) No.

(c) Some controlled experiments – see if you can think of what to do.

2.7 (a) On the western edge of the North Pacific gyre.

(b) At surface practically zero; at 1000 m silicate about 6 mg dm^{-3}, phosphate about 300 μg dm^{-3}.

(c) Yes.

(d) Nutrients are being released from decomposing materials but are trapped below a thermocline.

(e) Nutrient concentrations at the surface would increase, so reducing the sharpness of the step in the curves near the surface.

(f) Plenty of light and nutrients.

(g) A lack of nutrients, which are trapped below the permanent thermocline, limits primary productivity.

2.8 For tropical waters refer to 2.7 (f) and (g). For temperate waters the key factor in the summer is probably availability of nutrients, in the winter light.

2.9 This diagram clearly shows the interplay between the availability of light and nutrients in determining primary production. In the estuary the water is very turbid which limits primary production despite high nutrient availability. On the outer shelf the water is clearer but nutrient concentration is low, leading to a reduction in primary production. Maximum primary production occurs at an intermediate point where both light and nutrients are readily available.

2.10 (a) In spring it is the increasing availability of sunlight and the onset of thermal stratification, which reduces turbulent mixing and causes the bloom. In the autumn it is the breakdown of stratification with its associated resupply of nutrients to the eutrophic zone which triggers the bloom.

(b) Reduction in sunlight and increased turbulent mixing which takes autotrophs below the critical depth.

2.11 (a) Northern Polar and tropical.

(b) Check your answer (which should be substantial) with your tutor.

2.12 Most of the primary production is consumed before it reaches the benthos.

2.13 (a) Highest fish yields occur in areas with the highest primary production, shortest food chains and highest ecological efficiency. Coastal areas produce more fish per unit primary production compared to upwelling areas because they are more accessible to fishing boats.

(b) Used in zooplankton respiration or lost in faeces and urine.

CHAPTER 3

3.1 (a) Look at the shapes of their bodies, mouths and presence/absence of light producing organs.

(b) Pelagic plankton feeders feed low down in the food chain therefore expect them to have greatest biomass with demersal fishes, pelagic predators and vertical migrators coming next. Deep sea fishes have very low rates of food supply so they will have lowest biomass.

(c) Pelagic plankton feeders, but since they are small humans tend to eat demersal fish and pelagic predators.

3.2 (a) Expect the different stocks to become genetically different.

(b) Differences in spawning time of different stocks may be genetically determined.

3.3 You should draw a triangular diagram with adult feeding grounds, spawning grounds and juvenile feeding grounds as the three points of the triangle. Recruitment occurs when juveniles migrate to the adult feeding grounds. The arrow between the adult feeding and spawning grounds is a two way one. All the others are one way.

3.4 The plankton associated with either the spring or autumn blooms. It depends on spawning date.

3.5 (a) 1967–1968.

(b) The timing of the spring bloom.

(c) Development of the plaice eggs was delayed, because of cold sea water temperatures, so that the larval phase matched the spring bloom.

(d) Fish are genetically programmed to spawn at a particular time.

(e) Late January and early February, well before the spring bloom occurs.

3.6 (a) Should improve the match.

(b) Improve it.

(c) The sea is such a variable place and this variation in abiotic factors swamps the effects of increased stock size.

3.7 (a) Pacific.

(b) Atlantic.

(c) Industrialised countries with large, technologically advanced fishing fleets are located, primarily around the Atlantic and distance to the fishing grounds is shorter, hence total number of trips can be greater in the Atlantic.

(d) On the basis of the data supplied here, the Atlantic.

3.8 (a) The increased power of these ships meant they could pull larger trawl nets, therefore increasing efficiency.

(b) (i) Ensure smooth passage of the ground rope over the sea bottom and so prevent snagging. (ii) Keep the mouth of the net open.

3.9 (a) These fish are common and easy to catch, making such fisheries very economic.

(b) Use too much fuel and time to catch the fish, making such a fishery uneconomic.

CHAPTER 4

4.1 Competition for food, space, mates. Increasing predation, parasitism at higher densities.

4.2 $S_{n+1} = S_n + G + R - M$.

4.3 If biomass exceeds B_{max} predict that $G + R < M$, i.e. stock size declines back to B_{max}. *Vice versa* if stock size is less than B_{max}.

4.4 (a) $S_n = S_{n-1} + G + R - M - F$.
 (b) $G + R = M + F$.

4.5 (a) About 150 tonnes in each case.
 (b) 2000 tonnes, i.e. half B_{max}.
 (c) See answer to question 4.3.

4.6 (a) To avoid systematic bias in estimating population parameters.
 (b) Probably not since you are only sampling the fish you caught in your net not all the fish in the stock.

4.7 (a) 5446 fish.
 (b) 51%, i.e. $873/1720 \times 100$.
 (c) Multiply stock number by mean fish biomass.

4.8 (a) Fishing effort (f) is inversely proportional to stock size, fish become easier to catch as stock size increases.
 (b) It was decreasing in size; fishing effort (f) is increasing. Whilst catch per unit effort is decreasing.
 (c) Overfishing – the fishing effort is increasing.
 (d) It was decreased.
 (e) It decreased, i.e. fish were easier to catch.
 (f) Increasing.
 (g) Fishing effort continued to fall until 1944 with an increase in stock size. From 1944 to 1947 fishing effort increased slightly but with little change in c/f suggesting that this level of fishing effort produces a MSY.

4.9 (a) Harvesting rate is less than surplus yield so stock biomass would increase to B_1.
 (b) Harvesting rate now exceeds surplus yield, the stock becomes extinct.
 (c) If you get your estimates even slightly wrong the stock can become extinct or be irreparably damaged very rapidly.
 (d) Reduce your effort.
 (e) Profit.

4.10 (a) See text.
 (b) This goes against the classic concept of fixed effort fishing where overfishing is prevented by ensuring that catches vary as fish numbers vary – biologically sensible, but economically tough!
 (c) Economic pressure.

4.11 (a) 8
 (b) 1907
 (c) A good match between larval development and larval food supply.
 (d) (*i*) The age class is so strong that it may be tempting to set higher quotas and fish the stock harder in the belief that the good fishing will go on for ever.

(*ii*) Fishermen will have to accept that their catches will have to vary from year to year if they are not to overfish the stock.

CHAPTER 5

5.1 (a) The larger nets would only catch large fish. Currently there are only a few of these in the stock so it would lead to a short term reduction in yield. However, the measure would stop both recruitment and growth overfishing leading, eventually, to an increase in overall stock size.
 (b) Fishermen have to pay their loans and so are subject to short term economic, rather than long term biological, planning.

5.2 See section 5.1. Show your answers to your tutor.

5.3 (a) Fixed quota MSY.
 (b) Quota exceeded MSY as fishery grew and the MSY is variable from year to year because of climatic events like El Nino whilst the quota remains fixed from year to year.
 (c) Recruitment overfishing.
 (d) Economic pressures, the need for foreign currency, and social pressures, the need to provide jobs.
 (e) Use a fixed effort strategy though this would inevitably lead to a reduction in the number of people employed in the industry and a variable catch making financial forecasting, for example, more difficult. My bank manager would be unlikely to lend me money if my salary was to vary widely from year to year. The same rules apply to fishermen and whole countries.

5.4 (a) Before 0.6% of primary production went to humans. Now only 0.07%, i.e. a ten fold decrease.
 (b) Mainly gone into buried detritus.
 (c) Reduction in competition or perhaps anchoveta fry ate sardine eggs.

5.5 You should be able now to answer this question. Check and discuss your answers with your tutor.

5.6 Increase profitability. Think in terms of ecological efficiency, cost of production and harvesting.

5.7 The crowded conditions in the fish pens.

5.8 Unless Dichlorvos was specific to sea lice, which it is not, then local crustaceans, e.g. crabs, could also be affected.

CHAPTER 6

6.1 Chlorophyll absorbs blue and red light but not green.

6.2 (a) Only a tiny proportion of the energy trapped by plants is converted into animal production we can eat. The rest is lost as faeces, urine and in the process of respiration.
 (b) Where agriculturally valuable crops cannot be grown e.g. in deserts or mountains.

6.3 (a) To ensure that important changes in the instantaneous growth rates of plants are seen.
 (b) To ensure that the measurement is not biased by individual differences between plants.

6.4 (a) 0.009 g day^{-1}.

(b) RGR will fall.

(c) Because NAR measures changes in the efficiency with which a plant produces new biomass rather than changes in plant size *per se*.

6.5 This is the time of year when there is the maximum availability of sunlight.

6.6 (a) This is the result of ontogenetic drift.

(b) The result of decreasing availability of sunlight and ontogenetic drift.

(c) Because the curve produced for plants grown in the controlled environment of the growth room shows little change compared to the curve for the plants grown out of doors.

6.7 (a) Leaves lower in the canopy may be below the compensation point.

(b) Maximum LAI is likely to be reached late in the growing season when light availability may be quite low, leading to a fall in NAR; shading of the lower leaves; ontogenetic drift.

(c) More light will reach lower leaves whilst leaves high up in the canopy will still be photosynthesising at the maximum rate.

(d) Because crop growth rate depends upon the amount of leaves being displayed and the photosynthetic efficiency of those leaves.

6.8 Growth form of *daffodils* means plant close together and get higher NAR and higher LAI.

6.9 (a) Short stemmed varieties; the plant is putting more energy into producing grain and less into producing straw, the stem.

(b) Selective breeding has produced shorter stemmed plants which have higher harvest indices, though Norman is an exception to this compared to Hobbit. Clearly Norman has other characteristics, perhaps disease resistance, which increases its yield.

(c) Economic reasons – bread making wheats have lower HIs but attract higher prices; biological reasons e.g. disease resistance.

CHAPTER 7

7.1 Competition for light; competition for water; competition for nutrients; losses to pests in the field; losses to pests during storage; additional climatic factors e.g. wind; excess water leading to waterlogging and so on. Discuss your list and classification with your tutor.

7.2 (a) Both curves increase towards the middle of the year as both light intensity and light duration increase.

(b) The atmosphere contains a lot of water which warms up and cools down slowly – hence the lag.

(c) Follow same general pattern but lagging behind air temperature for same reason – soil contains a lot of water.

7.3 Light intensity and temperatures are greater on south facing slopes which lie north of the river (draw a picture if you cannot see the point).

7.4 (a) See text.

(b) Light intensity varies more the further you are away from the equator, which directly affects temperature whilst the sea has a moderating effect on climate.

(c) The length of the growing season will be longer in Britain than in the middle of the USA.

(d) Maize requires high temperatures for rapid growth and seed maturation. These are found in the USA but not Britain though some strains of maize can now be grown succesfully in southern Britain.

7.5 (a) All three are adaptations which enable plants to cope with life in a seasonal climate.

(b) Advantageous with crops which can overwinter and so resume growth earlier the following year, so making maximum use of available sunlight. Nuisance when it prevents farmers from sowing crops like sugar beet too early so that the plants make maximum use of available sunlight.

(c) Sow early so that soil temperatures are still high enough to ensure rapid germination. Acclimatisation.

7.6 By varying sowing date.

7.7 (a) Tropical = a; temperate = b; most northerly = c.

(b) See answer 7.2(a).

7.8 (a) Winter wheat since it achieves its maximum LAI and maximum NAR at about the same time leading to maximal CGR.

(b) Wheat because of the more erect growth of the plant's leaves.

(c) One where the early leaves were at right angles to the sun's rays but the later leaves grew more erect and so ensuring that the maximum amount of sunlight was intercepted when the canopy was poorly developed, but lower leaves were not shaded later on in growth.

(d) Plant seeds earlier so achieving higher LAIs when the maximum amount of sunlight is available.

(e) Maize plants have to be planted further apart than wheat plants.

7.9 (a) Exposing sugar beet to cold will vernalise it, so causing it to flower in its first year and reducing the quality of the root.

(b) Crop reaches maximum LAI earlier so making better use of sunlight.

(c) The cost of labour needed to transplant seedlings and the fuel costs of heating greenhouses make it an uneconomic proposition.

7.10 Intercropping is labour intensive which is expensive in developed countries.

7.11 (a) See text.

(b) Because (*i*) there is a greater diffusion gradient for water on a dry day compared to a humid one and (*ii*) the wind reduces the boundary layer around the clothing so increasing the rate of diffusion further.

(c) (*i*) Reduced as there is a reduced diffusion gradient from leaf to air (*ii*) increased – warm air can hold more water than cold air, so there will be an increased diffusion gradient from leaf to air. If temperature is too high photosynthesis will be reduced as plants close their stomata to conserve water.

(d) Boundary layer thickness is reduced which allows faster diffusion of CO_2 in to the leaf whilst diffusion of water from the leaf is not sufficiently high to cause stomatal closure.

7.12 The second statement is true but it does not provide an adequate explanation since it does not take into account factors like reduced lodging, temperature and relative humidity.

CHAPTER 8

8.1 (a) Acid soils.
(b) All essential nutrients available.
(c) Iron more available in acid soils.
(d) Acid ones.
(e) H^+ compete with Al^{3+} for cation exchange sites. Increasing concentration of H^+, i.e. lowering pH, will cause more Al^{3+} ions to come into the soil solution as H^+ dislodges them from the cation exchange sites.

8.2 Such a situation will promote waterlogging and the anoxic conditions in which hydrogen sulphide producing anaerobic bacteria grow.

8.3 By growing plants in a range of culture solutions which contained differing amounts of nutrients.

8.4 See text.

8.5 (a) Tropical soils contain less organic matter because of faster rates of microbial decomposition so they will be a lighter colour compared to temperate soils.
(b) Nutrients released by microbial decomposition in the soil are rapidly absorbed by tree roots so that nutrients become locked up in the trees.
(c) Soils formed where rates of decomposition are very low, e.g. on mountains or in polar regions where temperatures are low, or in waterlogged conditions where there is little oxygen and soil will be almost pure organic matter (peat) and hence very black.
(d) As a source of organic matter and as a mulch.

8.6 (a) The hydrogen bonding between water molecules.
(b) Increases the number of soil aggregates so increasing the number of spaces which can hold water.

8.7 (a) See text and Fig. 8.10.
(b) Available water is that water which is held in pore spaces which are sufficiently large for plant roots to extract the water from them. Unavailable water is the rest.
(c) No – see Fig. 8.10.
(d) This is water held in very small spaces where capillary action is so strong that the water cannot leave the space except when it is turned to vapour.

8.8 (a) Increases CO_2 content and decreases O_2 content through respiration.
(b) Plants take up nutrients by active transport from the soil solution. This requires a lot of ATP which is made most efficiently by aerobic respiration and so requires a source of oxygen close to the roots to be efficient.
(c) Cause waterlogging and anoxic conditions. Reduce production of ATP, reduce nutrient uptake and promote the growth of anaerobic bacteria which produce poisonous metabolic byproducts like hydrogen sulphide.

8.9 (a) It will contain more large soil aggregates and so more air spaces.
(b) Their high water content causes them to warm up slowly because water has such a high heat capacity. This means germination will take longer in these soils so that plants will not reach maximum LAI until late in the year.
(c) Prevents excess water getting into the soil whilst warming the soil up due to a mini greenhouse effect – both cloches and polythene let infrared rays in but not out.

8.10 See answer to Analysis: The water holding capacity of soils.

CHAPTER 9

9.1 Ploughing will uproot and bury existing weeds but will bring weed seeds to the surface to germinate.

9.2 Beneficial when the soil is not too wet and especially when ploughing occurs in the autumn so that frost can shatter large clods to form a fine tilth. See text for the problems caused by ploughing wet clay soils.

9.3 (a) Ploughing increases the aeration of the soil so increasing the rate of oxidation of organic material by microbial respiration. Earthworms eat this organic material. Reduce the rate of decomposition by not ploughing and you will increase the number of earthworms.
(b) Earthworms are nature's 'ploughs'. Their activity improves soil structure and tilth.
(c) The soil is kept covered either by a crop or crop residue.

9.4 (a) They will contain more water and have a smaller surface area compared to a ploughed soil.
(b) Have little effect on winter cereal growth but could limit germination of spring sown cereals which are therefore better grown by traditional methods.

9.5 (a) Check the soil pH.
(b) Deficiency of nutrients like nitrogen at high pH.
(c) Aids the decomposition of the grass and means that only one pass of a tractor is needed to complete two operations.

9.6 Heavy liming prior to planting of brassicas, with perhaps a little lime before the planting of legumes. No lime after root crops.

9.7 (a) Excessive nitrogen application makes the soil acid so this needs to be counteracted by the application of lime.
(b) Needed every year, usually at least twice in a growing season.
(c) Local watercourses or aquifers.
(d) Nitrogen is needed to make protein so a deficiency of fertiliser at this time would reduce the quality and hence the price of the grain.

9.8 See text.

9.9 (a) Potassium is not particularly mobile and there is still plenty of potassium in the cation exchange sites of the treated soil.
(b) Infrequent applications are needed on most soils.

9.10 (a) In the case of a nitrate fertiliser it could lead to burning but phosphate is so immobile in most soils that applying it close to the roots improves its availability.

(b) The concentration of nitrate in the soil solution will rise dramatically as the soil dries out leading to 'burning'.

9.11 (a) See text. Notice that to answer the question you need to find similarities and differences e.g. both raise nutrient levels in the soil but only organic fertilisers break-down to give humus which improves soil structure.

(b) To reduce the amount of carbon it contains whilst increasing the proportion of nitrogen.

9.12 Heavy metals are kept in solution in acid soils so that plants can take them up. The heavy metals can then damage plants and get into the human food chain through milk for example.

9.13 (a) Increases the rate of movement of water through the soil, water which will contain nitrates.

(b) Salinisation occurs when the soil solution evaporates from the surface bringing fresh soil solution, rich in dissolved salts, up into the surface layers of the soil by capillary action. Drainage, by removing excess water, prevents this.

(c) By lowering soil water levels the soil will warm up more rapidly in the spring.

9.14 (a) See text.
(b) Reduces soil erosion.
(c) One of the problems caused by moisture stress is lack of nutrients. Applying extra fertiliser will help to combat this.

CHAPTER 10

10.1 You might have – fast reproductive rate, good competitive ability, rapid dispersal, alternative hosts, ability to survive adverse conditions through dormancy for example.

10.2 (a) See text.
(b) It is a waste of money.

10.3 (a) Reduces ease with which a pest can spread from host to host.
(b) Cost – it is very labour intensive.

10.4 See text.

10.5 One year's seeding will produce an enormous seed bank which will provide seeds for germination over many years.

10.6 Old crops or their residues can provide alternative hosts for plant diseases.

10.7 Chewing insects tend to reduce photosynthetic area or physically weaken stems and roots. Insects with piercing mouthparts reduce the amount of photosynthate going to the plant and, more importantly, transmit viral diseases.

10.8 The repeated growing of potatoes on the same piece of land allowed the nematodes to feed and reproduce every year. Crop rotation breaks this cycle by planting crops the nematodes cannot feed on. Nematode numbers decline, therefore, to acceptable levels before the next potato crop is planted.

10.9 See text.

10.10 By providing crop residues from which the pathogen can infect new plants.

CHAPTER 11

11.1 (a) Weeds and crop residues are not buried. These provide alternative hosts for pests and pathogens which can then reinfect the crop plant.

(b) By the intensive use of pesticides, particularly herbicides.

11.2 (a) By planting crops at different times of the year; planting the same crop in widely separated, different parts of the farm in different years, i.e. crop rotation.

(b) Advantages reduce input of expensive pesticides; disadvantage is that you cannot grow the commercially valuable crops that you want to grow year after year to make the biggest profit.

11.3 See text.

11.4 Direct through competition for light, water nutrients and allelopathy (give an example). Indirectly by providing alternative hosts (give examples) for animal pests and plant pathogens.

11.5 (a) Broad spectrum – kills a wide range of insect pests, e.g. DDT; narrow spectrum – kills only a few, or even just one type of insect, e.g. Primicarb.
(b) See text, and discuss with your tutor.
(c) Persistent since carrot root fly can attack carrots at any time during the growing season. See text for definitions.
(d) Because you may want to eat the lettuce soon after applying the pesticide to the leaves. They have to be used continually to provide protection and that increases costs.

11.6 There are structural differences in the acetylcholinesterase of the two insects, due to genetic differences between them, so that primicarb binds to aphid acetylcholinesterase but not that of ladybird.

11.7 (a) No problems with drift, all the chemical gets to the crop.
(b) Aldicarb dissolves in the soil solution, is absorbed by the plant's roots and then translocated throughout the plant.
(c) Suggests that nematodes which come into contact with a solution of aldicarb in the soil, probably ingest it or absorb it through their 'skin'. It affects similar enzyme systems in different groups of animals.
(d) Plants like cotton are damaged by carbaryl.

11.8 Discuss your answers with your tutor.

11.9 See text and discuss with your tutor. Again look for similarities and differences.

11.10 (a) The control agent can be maintained at high densities under controlled conditions.

(b) The agent would itself then become extinct leaving the system to be reinvaded by the pest. The whole point of biological control is to keep the pest below a threshold.

11.11 Benefits include its ecological soundness, minimal use of pesticides and its eventual cheapness. Major costs are the in depth investigations needed at the outset, the education of farmers and the continual monitoring.

CHAPTER 12

12.1 The key criteria is that of harm or damage. Is the sewage causing damage?

12.2 No in all three cases.

12.3 Nickel = 150 ppb, 600 ppb, 600 ppb, 400 ppb.
Lead = 75 ppb, 60 ppb, 750 ppb, 500 ppb.
Manganese = 0.15 ppm, 1.5 ppm, 1.5 ppm, 1.0 ppm.

12.4 Point – easier to monitor a point source and so control it.

12.5 (a) High water velocity will enable the river to keep many particles in suspension, so increasing its turbidity.

(b) Clay particles are very small and so will remain in suspension in even quite slow flowing rivers.

12.6 (a) The treated effluent will have a low BOD anyway and the discharge will be further diluted and washed rapidly away by the river.

(b) The degree of dilution afforded by the small stream is much less, leading to a higher BOD in the water body. This depletes oxygen, resulting in anoxic conditions under which the anaerobic bacteria that produce hydrogen sulphide can thrive.

12.7 (a) See text.

(b) The pollutant can be transformed into innumerable compounds which the scientist then has to monitor and check. The likelihood is that some will be missed.

CHAPTER 13

13.1 (a) See text.
(b) 40.
(c) The slaughterhouse, probably because of the easily decomposed organic waste like blood.

13.2 The pond stratifies in the summer so that the bottom of the pond becomes anoxic, promoting the growth of the anaerobic bacteria which produce hydrogen sulphide. The stratification breaks down in the winter, allowing oxygen-rich cold water to reach the bottom of the pond.

13.3 (a) The river would become anoxic so that species which required oxygen in large quantities, or which did not have special adaptations to cope with low oxygen concentrations, would die.

(b) By increasing the rate at which oxygen dissolves in the water – see text for further details.

13.4 (a) This distance represents the time taken for the effluent to start decomposing rapidly.

(b) The sag would become more pronounced.

(c) The rate of oxygen consumption falls below the rate of oxygen supply.

(d) (*i*) A further sag would occur as the new sewage breaks down. (*ii*) An oxygen sag would occur as hot water can dissolve less oxygen than cold water, so once again the rate of oxygen consumption would exceed the rate of oxygen supply. (*iii*) The effect of this is to concentrate the effluent, thereby raising the water body's BOD. So another sag would probably occur, though it would depend on the amount of water removed.

13.5 (a) Because the diluting effects of lakes are less than that of rivers since the effluent is not carried away in a lake.

(b) To increase the rate at which oxygen dissolves in the water, thus ensuring that oxygen supply exceeded oxygen demand.

(c) Less water, slower flowing water and warmer water.

13.6 All to do with decomposition. BOD rises and then falls as the organic material is decomposed. This releases ammonia and nitrate, primarily as a result of the decomposition of proteins. Ammonia is released first and then nitrate, either by direct decomposition or by bacterial activity converting ammonia into nitrate. The initial fall in nitrate can be accounted for through uptake by rapidly growing bacterial populations. Levels of nitrate and ammonia fall as the amount of organic material available for decomposition falls and as the ions are absorbed by plants, e.g. *Cladophora*, and bacteria.

13.7 Dilution.

13.8 (a) See text.

(b) (*i*) You should have plotted distance on the *x*-axis and chemical variable on the *y*-axis. Check with your tutor to ensure you have chosen an appropriate scale. (*ii*) A is probably oxygen – a typical oxygen sag curve so you will need to explain the origins of this shape. B is probably BOD – again check with text to provide an explanation. (*iii*) You could have chosen any of the variables from Fig. 13.5.

(c) Size of the river, speed of water flow, presence of other pollutants e.g hot water.

CHAPTER 14

14.1 Enzymes only function effectively within a narrow pH range. Slight changes in external conditions can be temporarily damped out by an organism's homeostatic mechanisms. However, prolonged exposure or exposure to extreme environmental conditions, e.g. very acid water, will overwhelm these homeostatic systems.

14.2 Acidified soils around lakes will continue to leach H^+ ions into the lakes from cation exchange sites on soil particles since these soils contain so few bases which could neutralise the H^+ ions.

14.3 Snow melt will release a pulse of very acid water, severely disrupting the hatching of salmonid eggs.

14.4 Their gills will produce excess quantities of mucus, making diffusion of oxygen from water to blood more difficult.

14.5 (a) Under acid conditions nutrients like phosphate and nitrate become relatively insoluble.

(b) Decomposition is slow because microorganisms cannot thrive in the acid waters.

(c) Bases in the soil neutralise the acid rain.

14.6 Carp would thrive in the highly productive waters but swimming in turbid, algal rich water would be quite unpleasant.

14.7 This would return phosphorus from the hypolimnion to the epilimnion.

14.8 (a) During the day the algae will be above the compensation point and so produce more oxygen, by photosynthesis, than they consume by respiration. At night, however, plants will only respire and consume oxygen in the water.

(b) If the fluctuations are severe, so that night time oxygen levels are low, then fish may die.

(c) Because it was covered in ice preventing diffusion of oxygen from the air into the water.

(d) Organic pollution results in an increase in decomposers as they colonise the effluent and break it down, a process which consumes oxygen. Eutrophication involves the addition of excess nutrients, which stimulates the growth of producers. These blooms, however, eventually die and so provide an enormous input of organic material into the ecosystem which decomposers then break down, consuming oxygen in the process and so, potentially, producing anoxic conditions.

14.9 Discuss your answers with your tutor and friends.

CHAPTER 15

15.1 Sewage. See text.

15.2 (b) No, because drinking water is treated to remove bacteria.

15.3 (a) So that you only measure respiration in the bottle.

(b) If the material were highly refractory, i.e. difficult to break down. Wood pulp.

(c) and **(d)** You should find that you can easily distinguish between inorganic and organic sources of pollution on the basis of BOD – PV ratios. See text for explanation.

(e) The water becomes anoxic and foul smelling.

15.4 (a) It would not have one at all because no organism could live in pure, i.e. distilled, water!

(b) They contain haemoglobin which allows them to take up what little oxygen there is from the water and store it.

(c) No, only unpolluted by organic material.

(d) Both depend upon using frequency data, but biotic indices are really based on the concept of indicator species whilst diversity indices give all species equal weight.

15.5 They are rapid and could, potentially at least, be carried out by people with very little training.

Index

absolute growth rate 79
acid rain 168, 190–195
age classes, of fish 38–9
ageing fish 38
air temperature 94–5
anchoveta, Peruvian 62–5
anchovy 32–3
Antarctic circumpolar current 11
  upwelling in 14
aquaculture 67–72
arthropod pests 143–4
autotroph 5

bacterioplankton 5, 7
benthic division of marine
    habitats 2–3
benthic organisms 2
biochemical oxygen demand
    (**BOD**) 169, 179–80, 183–4, 206–8
biodegradation of pollutants 174
biological control 159–60
biological monitoring of
    pollution 210–15
biomagnification of pollutants 175–6
biotic indices 211–13
biotransformation of pollutants 175
'blue green algae' see Cyanobacteria

catch per unit effort 53, 54–6
catchable stock 37
cation exchange in soils 107–8
chemoautotroph 5
chemical oxygen demand (**COD**) 181,
    208–10
chemical transformation of
    pollutants 174–5
climate and plant productivity 91–6
cod 33–4
    Arcto-Norwegian stock 35–6
    North Sea stock 65–6
compensation depth 20–1
compensation point 17
copepods 6–7
critical depth 24–6
crop growth rate (**CGR**) 84, 101
crop productivity 94–8
    and light 99–102
    and temperature 94–98
    and wind 103–5
currents 11–14
Cyanobacteria 6

Danish seine nets 43–4
demersal fish 9, 33
diatom 5–6
dinoflagellate 6
digestible energy 87
dilution of pollutants 173
diversity indices 213–15

drainage of farmland 132–3
drift netting 42–3

eastern boundary currents 12–13
ecological efficiency of food
    chains 29–30
economic damage threshold 137
economic injury level (**EIL**) 136–7
edaphic factors 106–18
Ekman transport 12
El Nino 63
equatorial currents 11
  and upwelling 13
euphotic zone 3
eutrophication 169, 195–9
evapotranspiration 92

fertilisers 125–9
    organic see manuring
    run off see eutrophication
field capacity 115
fish farming see aquaculture
fish life history 33–34
    migration 34–5
    population dynamics 37–40, 48–9
    stocks see stock
    types 32–3
    yields 41–2
fisheries data 51–3
    management 53–9
    models 47–59
    regulation 61–3
fishing areas 41–42
fishing effort 53, 54–9
    optimum 54
fishing methods 42–5
fixed effort harvesting 57–9
fixed quota harvesting 55–7
food chains, marine see marine food
    chains
food webs, marine see marine food
    webs
fungal diseases 144–6
fungicides 152–4

gross primary production (**GPP**) 17
growth efficiency 29
growth rings in otoliths 38
    in fish scales 38
gyres 11

hadal zone 3
haddock 33
    North Sea stock 65–7
harvest index 87
harvestable dry matter 86–7
harvestable protein 87
harvesting fish stocks 50–1, 55–9
hedges 103–5
herbicides 151–2
herring 7–8, 32–3, 34
    age classes 38–9
    overfishing of 60
    stocks 36

heterotrophic 5
holoplankton 5
hook and line fishing 42
humus 112–13

indicator community 184–7, 211
    species 184–7, 211
insecticides 154–9
integrated pest control 161–2
irrigation 133–5

krill 6

larval mortality of fish 37
    match-mismatch hypothesis 39–40
LC50 200–1
LD50 200–1
leaf area duration (**LAD**) 84–5
leaf area index (**LAI**) 81–4, 101
light and crop productivity 99–102
lime 122–4
limiting factors 90–1
littoral zone 2–3

manuring 129–2
marine food chains 7–10, 29
    food webs 7–10, 64
    habitats 1–4
    nutrient cycles 9–10, 22–4
    productivity 17–29
        global pattern 18–19
        and light 20–22
        and nutrients 22–3
        and turbulence 24–6
        seasonal variation in 27–8
    secondary production 18, 29
    snow 9

match-mismatch hypothesis 39–40
maximum economic return
    (**MER**) 58–9
maximum sustainable yield
    (**MSY**) 51–9
mercury 175
meroplankton 5
mesopelagic zone 3
microplankton 5–6
minimal cultivation 121–2

net assimilation rate (**NAR**) 80–4, 101
nanoplankton 6
nekton 5
neritic province of marine
    habitats 2–3
neuston 5
North Sea fisheries 65–7
nursery grounds 34
nutrient depletion in euphotic
    zone 23
nutrients and marine
    productivity 22–3, 27–8

oceanic province of marine
    habitats 2–3

organic pollution 179–89
otoliths 38
otter board 45
overfishing 54, 60–7
oxygen demand 179–83
oxygen sag curves 181–3

particulate organic matter (POM) 7
pelagic division of marine habitats 2, 3–4
pelagic organisms 2
pest control 148–64
  biological methods 159–60
  chemical methods 151–9
  cultural methods 148–50
  integrated 161–2
pests 136–47
Petersen's method for ageing fish 38
photoautotroph 5
photosynthetic efficiency 74–6
phytoplankton 4–5
picoplankton 6
plaice 3, 34–5, 40
  catch quotas 57
plankton 4–7
plant diseases 144–7
plant growth 76–9
plant nutrients 91, 107–8
ploughing see tillage
pollution, definition 166–8
  types 168–72
polychlorinated biphenyls (PCBs) 201–2
primary consumers 7
primary producers 7
primary production 8
purse seine nets 43

quotas 56–7

recruitment, to a fishery 34
relative growth rate (RGR) 79–80

salmon fish farming 68–70
secondary consumer 7
secondary production 8, 18
seine nets 43–44
sewage 177–8
  effects 183–9
scales 38
size frequency analysis 38
soil 106–18
  air 107, 114–5
  cation exchange 107–8
  chemical composition 107
  organic matter 112–13
  pH 108–10
  physical structure 106–7, 110–13
  porosity 113–14
  temperature 91, 95–6
  water 107, 114–15
spawning grounds 34
standing crop 18
stock, of fish 34, 36
  age structure 37–40
  catchable 37
  growth 48–9
  recruitment to 34
  size 48–9, 53–9
    estimation of 51–3
surplus yield 50–51
  curves 51–9
  management models 54–59
suspended solids 169, 184

tagging of fish 52–3

temperature and crop productivity 94–8
temperature profile, of water column 14–15
thermal pollution 170
thermal stratification 15
thermocline 14–15
tillage 120–1
toxic pollution 199–202
toxicity testing 200–1
trawling 44–5
Trent Biotic Index (TBI) 211–13
tuna 32–5, 54–5

underfishing 54, 56
unit leaf rate (ULR) see net assimilation rate
unit stock 35–7
upwelling 12–14

Vigneron-Dahl gear 45
viral diseases of plants 146–7

Waddensee 34
water quality testing 204–6
waves 11
weeds 139–43
  annual 141
  perennial 141–2
whales 6, 57
wilting point 115
wind and crop productivity 103–5

year class 37

zooplankton 5–7